D0956083

RED ROCKS COMMUNITY COLLEGE

U18960 032 119 3

13276 - 2

QL   Cognitive
785.5   processes of
P7   nonhuman primates.
C6
1971

MAR 15 '74

FEB 2 3 '96

APR 0 7 2001

MAIL-WELL

COMMUNITY COLLEGE
OF DENVER

RED ROCKS CAMPUS

Cognitive Processes of Nonhuman Primates

Contributors

Norman Geschwind
John P. Gluck
Lee W. Gregg
Harry F. Harlow
M. K. Harlow
Leonard E. Jarrard
Donald R. Meyer
Raymond C. Miles
D. J. Mohr
Samuel L. Moise
David Premack
K. A. Schiltz
L. Weiskrantz

Cognitive Processes of Nonhuman Primates

Edited by

Leonard E. Jarrard

Department of Psychology
Carnegie-Mellon University
Pittsburgh, Pennsylvania

Academic Press   New York and London   1971

COMMUNITY COLLEGE OF DENVER
RED ROCKS CAMPUS

13276 - 2

QL
785.5
P7
C6
1971

202863

COMMUNITY COLLEGE OF DENVER
RED ROCKS CAMPUS

COPYRIGHT © 1971, BY ACADEMIC PRESS, INC.
ALL RIGHTS RESERVED
NO PART OF THIS BOOK MAY BE REPRODUCED IN ANY FORM,
BY PHOTOSTAT, MICROFILM, RETRIEVAL SYSTEM, OR ANY
OTHER MEANS, WITHOUT WRITTEN PERMISSION FROM
THE PUBLISHERS.

ACADEMIC PRESS, INC.
111 Fifth Avenue, New York, New York 10003

United Kingdom Edition published by
ACADEMIC PRESS, INC. (LONDON) LTD.
Berkeley Square House, London W1X 6BA

LIBRARY OF CONGRESS CATALOG CARD NUMBER: 77-154377

PRINTED IN THE UNITED STATES OF AMERICA

Contents

List of Contributors

Numbers in parentheses indicate the pages on which the authors' contributions begin.

NORMAN GESCHWIND (149), Harvard Medical School, and Neurological Unit, Boston City Hospital, Boston, Massachusetts

JOHN P. GLUCK (103), Regional Primate Research Center, Madison, Wisconsin

LEE W. GREGG (155), Carnegie-Mellon University, Pittsburgh, Pennsylvania

HARRY F. HARLOW (103, 121), Regional Primate Research Center, Madison, Wisconsin

M. K. HARLOW (121), Primate Laboratory and Department of Educational Psychology, University of Wisconsin, Madison, Wisconsin

LEONARD E. JARRARD (1), Department of Psychology, Carnegie-Mellon University, Pittsburgh, Pennsylvania

DONALD R. MEYER (83), Laboratory of Comparative and Physiological Psychology, The Ohio State University, Columbus, Ohio

RAYMOND C. MILES (165), University of Colorado, Boulder, Colorado

D. J. MOHR (121), University of Wisconsin, Madison, Wisconsin

SAMUEL L. MOISE (1), Department of Anatomy, University of California, Center for the Health Sciences, Los Angeles, California

DAVID PREMACK (47), Department of Psychology, University of California at Santa Barbara, Santa Barbara, California

K. A. SCHILTZ (121), University of Wisconsin, Madison, Wisconsin

L. WEISKRANTZ (25), Department of Experimental Psychology, University of Oxford, Oxford, England

Preface

This book is based on the Sixth Annual Symposium on Cognition that was held at Carnegie-Mellon University, March 26 and 27, 1970. Past symposia in the series have dealt with new approaches to and recent findings in the study of human thought (problem solving and concept formation, psycholinguistics, and cognitive processes in learning and memory). Participants focused on the current status of research dealing with complex behavioral processes of monkeys and apes.

Participants were encouraged to relate their research with nonhuman primates to relevant research dealing with cognitive processes of human primates. The extent to which this was accomplished is especially evident in the papers by Larry Weiskrantz, David Premack, Raymond Miles, and Norman Geschwind. As a result, many who attended the symposium left with a clearer understanding of the similarities and differences that exist in complex behavior of human and nonhuman primates.

The first day of the symposium was devoted to the presentation of papers comprising Chapters 1 through 4 and 6 of this book. In Chapter 1 by Jarrard and Moise the main concern is with short-term memory in the monkey and how this relates to human short-term memory. Larry Weiskrantz, in Chapter 2, is more concerned with comparing memory deficits that accompany brain dysfunction in animals and man. The analysis of the development of language in a young female chimpanzee is described by David Premack in Chapter 3. Don Meyer, in Chapter 4, presents a cogent analysis of how habits interact with concepts in the monkey. Although the material in Chapter 5 by Gluck and Harlow was not presented at the symposium, this review of the literature serves as an additional contribution to the book and background for the chapter that follows. In Chapter 6 Harry Harlow and his colleagues describe the recent results of their extensive program of research concerned with the effects of early deprived and enriched environment on later complex behavioral processes of monkeys.

The second day of the symposium was devoted to discussing the above papers from different points of view. Thus, Norman Geschwind, in Chapter 7, comments on several of the contributions using a neuroanatomical approach, and Lee Gregg, in Chapter 8, provides an interesting analysis of the papers from an information-processing point of view. Raymond Miles, in Chapter 9, discusses the papers that were concerned with memory and presents some of his recent important research comparing delayed-response performance of several species of monkeys, several age groups of children, and adult subjects. These discussions and the informal exchange that resulted added considerably to the success of the symposium, and the resulting chapters contribute significantly to the present volume.

I would like to express my appreciation to the participants for their cooperation and for their time and effort. Without them this volume would not exist. The symposium was supported in part by funds from Carnegie-Mellon University. I also gratefully acknowledge the very able assistance that was provided by Betty H. Boal, who handled many of the administrative and secretarial details involved in arranging for the symposium, and Barbara Gourley, who helped in preparing this volume. And, to my wife, Janet, I want to express my sincere appreciation for her understanding and support.

Leonard E. Jarrard

CHAPTER 1

Short-Term Memory in the Monkey[1]

Leonard E. Jarrard and Samuel L. Moise

Over recent years an increasing amount of interest in memory has been in evidence from studies on both human and animal levels. On the human level most empirical and theoretical contributions have centered around the relationship between short-term memory (STM) and long-term memory (LTM) (Atkinson & Shiffrin, 1968; Norman, 1969). Most recent studies with animals have been concerned with memory consolidation and the underlying neural and chemical bases of memory (Lewis, 1969; Deutsch, 1969). Formal attempts to assess differences in memory between human and nonhuman species have been few. However, if we are going to truly understand memory, it is important that studies be undertaken to determine the similarities and differences that exist between man and those close to him on the phylogenetic tree.

Perhaps the most reasonable place to begin in attempts to integrate human and animal memory would be to investigate STM as it exists across species, especially as it exists for man and his closest phyletic relatives, the monkeys and apes. The development of new techniques for investigating human STM have resulted in many important empirical and theoretical contributions (Melton, 1963; Norman, 1969). Comparable research does not exist for the nonhuman primates. It is of historical interest that early comparative psychologists attempted to attack the "higher mental processes" in monkeys and apes by employing the delayed-response procedure (Harlow, 1951). Delayed response was felt to involve the central representation of a stimulus that was not physically present. More recently there have been many studies in which delayed-response and delayed-alternation tasks have been used (Fletcher, 1965), but this research has not been concerned with determining STM as it is usually studied in the

[1] This research was supported by grants from the National Science Foundation (GB-8044) and United States Public Health Service (MH-07722).

human. In order to obtain the kind of STM data for monkeys that could be compared with the human STM literature, it is first necessary to design more comparable testing procedures. It should be useful at this point to consider briefly approaches used in studying human memory.

The distinction between STM and LTM is not always clear and has been a subject of considerable controversy (Adams, 1967; Melton, 1963). Generally, the study of LTM involves employing multiple trials with supraspan lists of items and retention being measured over days. Research in STM characteristically utilizes subspan amounts of material presented once with retention being measured over seconds. The evidence supporting the STM-LTM distinction has been used by several investigators to develop two-process (Waugh & Norman, 1965) and three-process (Atkinson & Shiffrin, 1968) models of memory. These models postulate a limited capacity short-term storage mechanism (memory span) with a rehearsal loop that renews the strength of material in short-term storage to help it enter long-term storage. Other investigators argue that similar laws apply to both sets of data and a STM-LTM distinction is therefore unnecessary (Melton, 1963). This issue need not concern us here. It should be sufficient to say that we are interested in STM, operationally defined as the retention of events over a period of a few seconds.

Probably the most useful experimental procedure devised to study human STM is the one first employed by Peterson and Peterson (1959). These investigators determined the retention of single trigrams after intervals of from 3 to 18 sec. By having the subjects count backward by 3's or 4's during the delay interval between presentation of the material to be remembered and the test for recall, there was complete forgetting after intervals of only 18 sec. The procedure of counting backward was assumed to prevent, or make more difficult, rehearsal of the to-be-remembered items. If rehearsal is not prevented, the event is encoded and becomes available for long-term storage (Slamecka, 1967). Thus, retention in the Petersons' situation possibly reflects strength of the memory trace free from the confounding effects of rehearsal or practice. Naturally, it is not known if the processes assumed by the terms rehearsal and encoding have analogs in the nonhuman primate. However, by employing procedures similar to those used in human studies, we should be able to see if comparable data are obtained for monkey and man. We would then be in a better position to talk about underlying processes in the monkey.

One must, of course, be extremely careful in attempting to generalize across species since the laws governing memory of a particular animal may be species-specific. In trying to understand similarities and differences in STM of monkey and man, it seems that two rather different research strategies can be employed. One approach would be to determine the laws, principles, or regularities that govern STM in the monkey, and then systematically devise a process theory to account for the data. By employing a process approach, it would eventually be possible to compare the resulting theory of monkey STM with the existing human STM theories. A second strategy, one that can be referred to as the comparative approach, would be to compare monkey and man directly and concentrate on *differences* in STM. One of the main problems with this approach is in developing comparable situations for the two species that are free from unwanted confounding variables. Of the two approaches, it seems that developing separate theoretical frameworks will eventually permit more meaningful comparisons and provide the better vehicle for understanding STM as it exists across species. Unfortunately, the kinds of data one needs for a theory of STM for the monkey does not exist. Therefore, as a first step it seems important to demonstrate that STM can be studied in the monkey using procedures like those employed in the study of human STM. In the present research, a procedure similar to the one devised by the Petersons to study retention in humans has been adapted to study STM in the monkey.

In this paper we summarize a series of studies where several of the more important variables in human STM are investigated in the monkey. The experiments to be described are as follows: (1) the effects of physical or overt restraint on delayed-response performance; (2) the role of interpolated activity in STM; (3) the importance of repetition or practice of the stimulus to be remembered; (4) level of difficulty of stimulus material and how this affects retention; and (5) intertrial interval effects on STM.

THE TASK

A delayed-matching-to-sample task (DMS) (Blough, 1959; Jarrard & Moise, 1970; Scheckle, 1965) was used in the present research to study STM in the monkey. It would seem that this task is preferred to delayed response and delayed alternation since the possibility always exists in spatial tasks that the delays are "bridged" with the help of bodily orientation (Fletcher, 1965). Even though one can never be certain that motor responses are not "bridging the gap" in DMS, this possibility is minimized. Another reason for employing this task is that it readily lends itself to precise control by the experimenter. Variables such as stimulus duration, number of stimulus presentations, delay intervals, and interpolated tasks can be carefully controlled and precise response information obtained.

As used in the experiments to be described, matching to sample consisted of the presentation of a sample stimulus followed by the presentation of two test stimuli, one of which was similar to the sample. The subject's task was to choose the test stimulus that matches the sample. In DMS performance a delay interval was introduced by varying the time between termination of the sample and presentation of the test stimuli.

GENERAL PROCEDURES

Since many of the general aspects of the procedure are the same throughout the series of studies, these will be described before reviewing the experimental findings. The animals were trained and tested in a computer-controlled system that has been described in detail elsewhere (Moise & Jarrard, 1969). The computer was a 16-bit DDP-116 (Honeywell), with 16 K core storage, located in the computer-controlled psychology laboratory. The computer was used on-line to select and present stimuli randomly, to carry out all timing functions, to deliver reinforcements, to determine when criteria for position habits and learning are met, etc. In addition, complete response information was recorded.

The computer was interfaced with an environmental chamber in which the animals were tested. The test panel had three circular display holes (1-in. diam) arranged in a triangle with the apex at the top and a rectangular pellet feeder opening located below the base of the triangle. Behind each circular opening was a one-plane digital readout projection unit that permitted presentation of 12 different stimuli. Between the readouts and the panel were translucent plastic response manipulanda hinged to microswitches. Correct responses were rewarded in the first two experiments with 190 mg Ciba banana pellets and in the other experiments with soybeans. The test chamber was designed so the animals could be tested either while free to move around the compartment or while restrained in a removable primate chair.

In training subjects to perform the DMS task, the monkeys were first taught to press any one of the three lighted displays (white light). They then progressed through a series of preliminary training programs consisting of discrimination training, simultaneous matching from sample with correction, simultaneous matching from sample without correction, and DMS with titrating delays starting at 0 sec and increasing in the sequence ¼, ½, and 1 sec (see Moise & Jarrard, 1969, for a more complete description of the apparatus and preliminary training procedures). After successfully performing DMS at short delays, the subjects were ready for experimental studies with longer delay intervals.

EFFECTS OF PHYSICAL RESTRAINT OF
BEHAVIOR ON SHORT-TERM RETENTION

The essence of the Peterson and Peterson (1959) experimental paradigm is control of behavior during the delay interval between presentation of the stimulus material to be remembered and the test for retention. There have been several attempts to investigate the importance of type of activity in which monkeys engage during the delay in delayed-response performance. However, attempts to actually control behavior during the interval have been few. Malmo (1942) carefully observed activity of monkeys during the delay interval and found they performed better when darkness was maintained during the delay than when the delay period was filled with continuous light. Other investigators (Pribram, 1950; Weiskrantz, Gross, & Baltzer, 1965) reported that central nervous system depressants improve performance of frontal-lesioned monkeys on delayed-response tasks. Presumably, the effect of these manipulations was to decrease activity during the delays and/or change sensory input.

Early investigators of delayed response were concerned with the possibility that animals might "bridge the interval" by maintaining body postures or orienting responses throughout the delay (Hunter, 1913). Although it has been reported that some monkeys do utilize bodily positioning or visual orientation (Miles, 1957), Gleitman, Wilson, Herman, and Rescorla (1963) have shown that solution of the spatial delayed-response problem does not depend on such responses. In an interesting study, Nissen, Carpenter, and Cowles (1936) found a facilitation of delayed-response performance in chimpanzees when the animals were required to move physically to the correct side of the cage before initiation of the delay. From the existing literature, it would seem that orienting responses are not necessary for successful delayed-response performance in primates.

Observations in our laboratory indicated that monkeys often engaged in many behaviors during delays in DMS performance that would appear to interfere with successful performance (i.e., grooming, moving around the testing compartment, looking into a one-way glass observation window). Our first experiment was designed to determine the importance of these hypothesized "incompatible" responses by systematically comparing performance when subjects were free to move about the compartment with performance when subjects were physically restrained in a primate chair (Jarrard & Moise, 1970). Since in the restrained condition the subjects were unable to move from in front of the stimulus display, it was expected that they would attend better, there would be less motor activity during delay intervals, and the resulting performance would be better than in the nonrestrained condition. A second purpose of the study was to decide whether in future STM experiments the subjects should be tested while restrained or while free to move about the testing compartment.

The subjects in this study were one female and two male stumptail macaques (*Macaca arctoides*) weighing 4.9, 4.3, and 3.8 kg. Following preliminary training the subjects were tested in DMS with a titrating schedule of delays, that is, correct responses served to increase and incorrect responses served to decrease the delay intervals. Each session began by having the subject match to sample with a 0-sec delay between response to the sample and onset of the test stimuli. When the subject made three successive correct matching responses the delay increased by 2.5 sec, whereas two successive incorrect responses served to de-

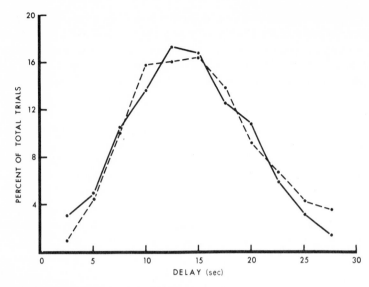

·FIG. 1. Percent of the total number of trials at each delay for restrained and nonrestrained conditions. (- - -) Restrained; (——) nonrestrained. (From Jarrard & Moise, 1970.)

crease the delay by a similar amount. Each subject was given a total of 100 trials each day at an 8-sec intertrial interval. Testing was carried out both while the subjects were free to move around the compartment (nonrestrained condition) and while restrained in the primate chair (restrained condition). Performance had stabilized after 25 days of being exposed to the titrating delays. During the actual study the subjects were run under the restrained condition for 5 days and nonrestrained for 5 days for three replications of each. Thus, each subject was tested for 15 days in each of the two conditions.

The percent of the total number of trials at each delay generated by the subjects performance and the titrating procedure is shown in Fig. 1. Analysis of the data showed no real differences between performance in the restrained and nonrestrained conditions either for the different delays or for subjects. With this titrating procedure of setting the delays, the subjects most frequently worked at delays of 12.5 and 15.0 sec. Data on latencies of response following onset of sample and test stimuli indicated that latencies to sample stimuli were significantly longer in the nonrestrained as compared to the restrained condition but latencies to test stimuli did not differ. Generally, these data show that subjects performed as well while free to move about the testing compartment as while restrained in front of the test panel in the chair.

The STM curves plotted as percent correct for each delay interval are shown in Fig. 2. In addition to a composite curve, individual subject curves are presented to show individual differences. It is apparent that the group curve has the classic decay form found in human experiments when interpolated activity prevents rehearsal during the retention interval (Melton, 1963). Although subjects. did reach longer delays than those indicated in the individual curves, these data were not plotted since they represented fewer than 1% of the trials. An analysis of the data for all subjects showed that performance at delays of 27.5 sec differed significantly from chance but performance at 30.0 sec was not significantly different. It should be pointed out that the STM curve shown in Fig. 2 would differ in slope and maximum delay with different experimental conditions. Our subsequent research describes several of these conditions (see below).

It is clear from this experiment that the subjects performed the delayed-matching task equally as well while free to move about the testing compartment as while physically restrained in the primate chair. These results are surprising in view of the finding in studies with humans that intervening activity tends to interfere with retention in STM tasks (Waugh & Norman, 1965). Apparently, the motor responses in which the subjects characteristically engaged during the delays were not sufficiently demanding in terms of attention to interfere with

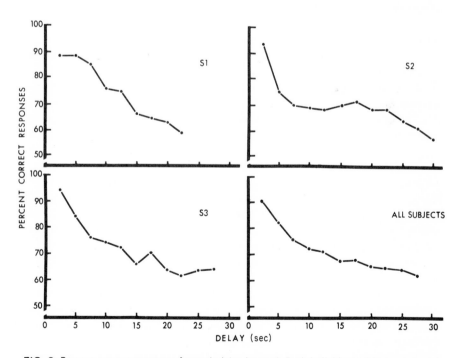

FIG. 2. Percent correct responses for each delay interval. Both individual subject curves and a composite curve are presented. (From Jarrard & Moise, 1970.)

STM. Because of these results, the monkeys were not restrained during subsequent testing.

THE ROLE OF INTERPOLATED ACTIVITY
IN SHORT-TERM RETENTION

The attempt in the next experiments was to go beyond physically limiting motor movements during the delay to bringing interval behavior under more rigid experimental control. In these studies, carried out in our laboratory by Samuel Moise (1970) as part of a dissertation, the monkeys were required to perform a motor activity during delay intervals. If STM in the monkey is similar to STM as it is studied in humans, one would predict that the imposing of a sufficiently difficult task during the delay should interfere with retention.

An important problem concerns the nature of the intervening activity. In human STM research subjects are most often required to carry out a difficult task such as counting backward (Peterson & Peterson, 1959) or listening to a list of words some of which must be recalled (Murdock, 1961). Williams, Beaver, Spence, and Rundell (1969) have shown that recall of digits by human subjects is only slightly disturbed when the interpolated task is a kinesthetic one (moving a lever to different angles). It is not known what kind of intervening task would be sufficiently difficult to use in studying interfering effects in monkeys. Since monkeys and apes are nonarticulate and depend for the most part on motor behavior, one would think that an interpolated motor activity would be a more effective disruptor of memory for these subjects than it is for humans. A difficult motor task incorporated during the delay might be expected to prevent or interfere with whatever processes are necessary for correct matching-to-sample performance. In the experiments to be described, a reaction-time task was employed. The subjects were required to respond to a white light that was presented during the delay period. The task was a rather difficult one since it was necessary for the subject to attend carefully to the stimuli and respond correctly within a short period of time following onset of the stimulus. Two important variables in STM were investigated in these studies—the amount of interpolated activity, and when during the delay interval the interpolated activity was required.

In the first experiment different amounts of interpolated activity were presented during each of several delays. It was assumed that increased interpolated activity would represent a more demanding or interfering processing load and the result would be a greater disruption of STM.

The subjects in these experiments were the same three stumptail macaques used in the previous study. Preliminary to the actual start of the experiment the subjects were adapted to random presentation of a fixed set of delays, followed by training designed to bring behavior during delay intervals under stimulus control. As a first step in training, responding to manipulanda during delays was extinguished by terminating the trial and turning off the house lights for 3 sec when such responses were made. The second stage of training consisted of reducing response latencies to onset of the white light. This was accomplished by using a titrating procedure to set the maximum allowable latency (two successive responses within the allowable time served to decrease and two incorrect responses served to increase the maximum response time). Each day the maximum allowable response time began at 3 sec and then decreased in the sequence 2.0, 1.5, and then in 0.1-sec steps down to 1.0 sec. If the subject did not respond within the allowable time, the house lights were turned off for 3 sec, and following the 15-sec intertrial interval, the next trial began. Average response latencies decreased over sessions from 1.32 sec on the first 2 days of training to 0.97 sec by the last 2 of the 11 training days.

Following reaction-time training, the subjects were required to respond to 1, 3, or 5 successive reaction times before being reinforced. The final stage of training consisted of presenting the reaction-time task during the delays in the DMS task. During the actual experiment subjects were required to respond to 0, 1, or 3 interpolated stimuli that were presented at the beginning of either 0-, 5-, 12-, 20-, or 30-sec delays. The first interpolated stimulus was presented 0.5 sec following the response to and termination of the sample. If the subject responded to the white light within 1.25 sec, the white light was terminated and the remainder of the reaction-time interval plus 0.5 sec between interpolated stimuli elapsed before presentation of the next stimulus. In the situation where the subject did not respond to an interpolated stimulus within the allotted 1.25 sec, or when he made a response during the delay when no interpolated stimulus was present, house lights were turned off for 3 sec and the next trial began after the 15-sec intertrial interval. The subjects received 65 trials each day.

Results of these experimental manipulations on performance are shown in Fig. 3 where percent correct responses are plotted as a function of delays and interpolated stimuli. It is obvious from inspection of the curves that interpolated activity interfered with STM. Further, the amount of interference depended on the number of interpolated stimuli that the subjects responded to during the delay interval. With no interpolated stimuli 74.5% of the trials were correct at the 30-sec delay, whereas with one and three interpolated stimuli performance was at 68.6 and 65.8%, respectively. Analysis of the data showed that even with three interpolated stimuli performance was still above chance at the 30-sec delays.

In the original Peterson and Peterson (1959) experiment where counting backward was used as interpolated activity, percent of the items correctly recalled was approximately 0% after delays of only 18 sec. Obviously, our experi-

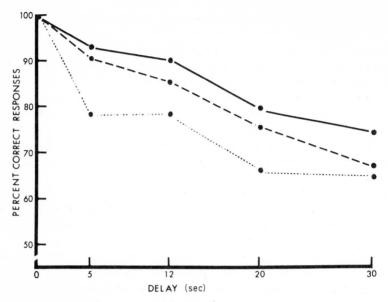

FIG. 3. Percent correct responses as a function of delays and interpolated stimuli (IS). (——) 0 IS; (- - -) 1 IS; (•••) 3 IS.

mental procedure differs in many ways from the Peterson and Peterson task. Perhaps one of the most obvious differences, as far as the present experiment is concerned, is that in the Petersons' situation the retention interval was completely filled with interpolated activity. From looking at the curves in Fig. 3, one might expect that rate of forgetting would increase if the monkeys were required to perform the interfering task continuously throughout the delay interval. The next study was designed to test this hypothesis. In addition, it would be interesting to compare the effects of an interfering activity carried out immediately after presentation of the to-be-remembered stimulus with interfering activity performed immediately before onset of the test stimuli. Certainly, when the interference is of a more severe kind (such as electroconvulsive shock), the period immediately after learning is most critical (Lewis, 1969). One might think of this as looking at the effects of a disrupting activity immediately after storage of stimulus information versus disruption immediately preceding retrieval of memory traces.

In the next study, either three stimuli were presented at the beginning of the delay interval, three stimuli were presented at the end of the delay, there were stimuli successively presented throughout the delay, or there were no interpolated stimuli. The same three monkeys were used with the four interpolated stimulus conditions and delays of 0, 5, 12, and 20 sec. A total of 110 trials were run each day for 14 days.

The effects of presenting interfering activity continuously and at different times during the delay interval can be seen in Fig. 4. The greatest disruption of STM was obtained when the reaction-time task was repeatedly presented throughout the retention interval. The effects of three interpolated stimuli at the beginning of delays were indistinguishable from three stimuli at the end, but both conditions were significantly different from zero interpolated activity. Generally, the overall level of performance was improved over that found in the previous study (see Fig. 3).

Short-term retention was affected by amount of interpolated activity. It does seem reasonable that increased interpolated activity would represent a more demanding or disrupting processing load and the result would be an interference with STM processes. Most of the studies with humans have confounded either rate of processing or length of retention with amount of interpolated activity (Murdock, 1967; Peterson, Saltzman, Hillner, & Land, 1962). However, Kulp (1967) systematically varied rate of interpolated activity at different retention intervals and concluded that the greater the processing demands on the subject, the more STM is affected. For monkeys, amount of an interpolated activity (and therefore amount of processing) carried out during the delay interval is also an important variable influencing STM performance.

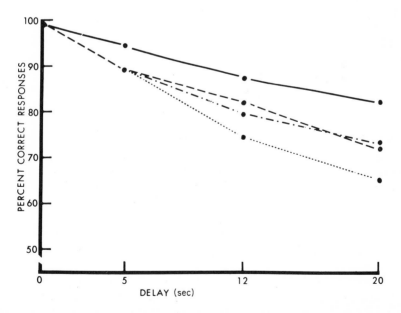

FIG. 4. Percent correct responses as a function of delays and interpolated stimuli (——) 0 stimuli; (- - -) 3 stimuli at the beginning of the delay; (• —•) 3 stimuli at the end of the delay; (• • •) continuous stimuli.

One might think that new material would be more susceptible to interference early after exposure than later when it has become better "organized" (Hebb, 1949). The point of interpolation of interfering activity was not important in the present situation. This finding suggests that interfering activity does not differentially affect consolidation and retrieval processes. Essentially the same findings were reported by Jarvik, Goldfarb, and Carley (1969) in a DMS task; however, Fletcher and Davis (1965) found greater disruption of delayed-response performance in a Wisconsin General Test Apparatus when an opaque screen was lowered early, as compared to later, in the delay period. In research with humans, Corman and Wickens (1968) found no relationship between amount of retention and temporal position of an intervening activity. However, Pylyshyn (1965) reported a greater decrement when interpolated activity took place early in the delay. While it is not possible to draw any definite conclusions concerning the effects of temporal position of intervening actvitity during delays on STM, it should be pointed out that temporal position was not important in the present studies even though the interpolated activity had an overall disrupting effect on STM.

These two interpolated activity studies are consistent in showing that STM is affected by an interfering activity carried out during the delay interval. Several studies have been reported where distracting stimuli were presented with no attempt to control behavior. Meyer and Harlow (1952) found that lowering an opaque screen during the delay interval in delayed-response performance increased errors by rhesus monkeys. More recently, Jarvik, Goldfarb, and Carley (1969) presented stimuli at different times during delays in a DMS task. It was reported that a negative stimulus (color different from the sample) resulted in a disruption of correct performance, a neutral stimulus (white light) produced little or no effect, and a positive stimulus (color same as the sample) facilitated performance. In both studies distracting stimuli were presented a relatively few number of times, and responses made by the subjects were not controlled. Jarvik suggested that with further testing his subjects would probably have habituated completely to the nonreinforcing interfering stimuli and as a result the stimuli would no longer affect delayed-response performance. Our studies differ in that interpolated activity was under strict stimulus control and some degree of cognitive processing was demanded of the subjects. The consistent results that were obtained over the two studies and several thousand trials would seem to argue against further experience eliminating the effects of the interpolated activity. This stability suggests to us that the results may reflect basic properties of STM in the monkey.

Latency of responses to test stimuli were not related to percent correct responses in any simple way. Percent correct performance decreased with increasing delays, whereas latency of responses tended to increase. These trends were found in all studies and are shown for the first study in Fig. 5. Latencies

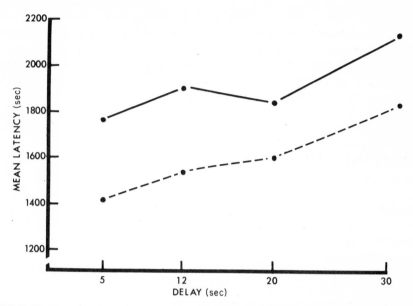

FIG. 5. Response latencies on correct (- - -) and incorrect (——) trials for each delay interval.

for correct responses were shorter than those for incorrect responses and differed in showing a significant linear increase over delays. Latencies for incorrect responses showed an increase with increasing delay intervals but no linear trend. It would appear that response latencies on correct trials and percent correct data are measuring different processes. In both experiments, latencies were found to decrease over days of testing while percent correct performance remained essentially the same. Results obtained in human learning studies have convinced several investigators that response latency may be a function of processes like attention and retrieval, while percent correct is a measure of strength of the memory trace (Murdock, 1968). A similar interpretation applied to the present data suggests that both strength of the memory trace and attention decreased with increasing delays.

A rather surprising result of these studies with monkeys is that STM, even with continuous interrupting activity throughout the delay interval, was significantly above chance at 20-sec delays. This is in contrast with the near-chance recall of trigrams in human subjects after only 18 sec of interpolated activity. Although there are many differences in the two situations, one possibility is that the reaction-time task used in the present monkey studies was not demanding enough to produce sufficient disruption of the memory trace. Effectiveness of interpolated activity has been a problem in human STM research (Keppel, 1965) and would deserve further consideration in research with monkeys. Another important consideration is the nature of the stimuli used. By the end of these studies, the monkeys must have been extremely familiar with red and green stimuli. Level of difficulty and meaningfulness of material have been shown to affect slope of the STM curve for humans (Peterson, Peterson, & Miller, 1961; Pylyshyn, 1965) and would probably be important variables in monkey STM.

THE IMPORTANCE OF REPETITION OF
THE STIMULUS TO BE REMEMBERED

The number of repetitions of the to-be-remembered stimulus should be an important variable in monkey STM. It would be interesting to see if in the DMS situation one could affect the storage of a single unit of information (red or green stimulus) by presenting the stimulus different numbers of times before onset of the delay. One would predict that probability of retention at a given delay should increase as a function of repetitions of the stimulus.

There is evidence to support such a prediction from the human literature. Brown (1958) suggested that the effect of repetition or rehearsal of a stimulus would be to postpone the onset of decay rather than affect strength of the memory trace. He felt that forgetting curves would be similar regardless of the number of occurrences of the stimulus. In the original Peterson and Peterson (1959) study it was found that rehearsal did tend to increase the probability of recall and thus these investigators concluded that rehearsal affected strength of the memory trace. Subsequently, Hellyer (1962) systematically investigated the importance of repetition and found that with 1, 2, 4, and 8 repetitions of the to-be-remembered stimulus, retention increased accordingly.

In order to look at the importance of repetition of the stimulus in STM in the monkey, we used the DMS situation and programmed the computer to present the sample stimulus either one, two, or four times. Upon presentation of the sample for the first time, the subject had the usual 15 sec within which to respond. Once the sample was pressed, the light was terminated as usual, and in the two and four repetition conditions there was 0.5 sec of no light followed by presentation of the sample again. The subject had 1.5 sec within which to respond on each repetition of the sample. If no response was made within the allotted time, the house lights were turned off for 3 sec followed immediately by the 15-sec intertrial interval. The delay interval was measured from the response to the last sample in each trial. Delays of 0, 5, 15, 30, and 60 sec were orthogonally presented with 1, 2, and 4 repetitions of the sample for a total of 90 trials a day and 15 days. The subjects were the same three stumptail macaques previously described.

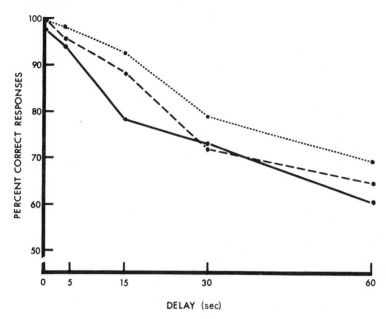

FIG. 6. Percent correct responses as a function of delays and number of repetitions of the to-be-remembered stimulus. Sample presses: (———) 1; (- - -) 2; (• • •) 4.

The results of this experiment are shown in Fig. 6. It should be noted that the effect of repeatedly presenting the sample was to increase STM. Although the interaction of repetition conditions by delays was not significant, there was an overall statistically significant difference between 1 and 2 repetitions and between 1, 2, and 4 repetitions. It is interesting to look at performance at 60-sec delays. This is the longest delay used in any of our studies. When there was one presentation of the sample, the subjects were correct on 60.4% of the trials, a level that is significantly different from chance. With 2 and 4 repetitions of the sample, above chance levels of 64.8 and 69.1% correct were obtained at 60-sec delays. It is apparent that repeatedly responding to the sample did facilitate retention.

An important question to ask about this experiment is whether retention was improved because the correct stimulus was repeatedly presented or whether the procedure of repeating the sample resulted in changes in attention or increases in strength of the orienting response. Repeated presentation of the sample may just increase the probability that the subject will see the sample and/or force him to attend more carefully to the stimulus. One way to test this attention hypothesis is to have the subjects respond to 0, 1, or 3 "neutral" white lights before responding to the red or green stimulus. Thus, in the equivalent to the four-press

condition of the previous experiment the subject would respond to three successive presentations of a white light followed by presentation of the correct stimulus. If performance improves with repetitions of the neutral stimulus, then the
observed increases in STM with repetition of the sample can be attributed to
increases in attention and not increases in strength of the memory trace. This
study was carried out after completion of the intertrial-interval experiment described below.

Figure 7 shows the results of this control study. Although performance was
improved over that in the previous study (as evidenced by comparing conditions
when the sample was presented once without preceeding stimuli), presses to one
and three white lights before presentation of the correct stimulus interfered with
retention. Performance in the one and three white light conditions did not
differ, but both of these conditions differed significantly from performance
when no white light preceeded the sample. It is interesting that the amount of
interference was as much with this proactive procedure as it was when the white
light stimuli were presented *after* the sample (see Fig. 3). The results of this
experiment suggest that the improved retention obtained when the sample is
repeatedly presented cannot be attributed to differences in attention.

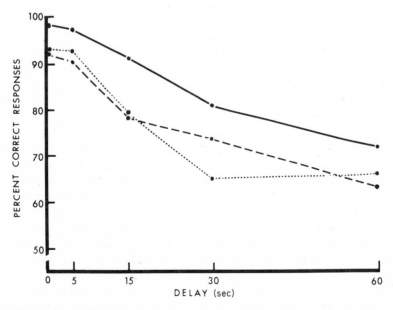

FIG. 7. Percent correct responses as a function of delays and number of repetitions of the
white light before presentation of the sample stimulus. White light presses: (——) 0; (- - -) 1;
(· · ·) 3.

It is encouraging that one can increase STM in DMS performance by having monkeys respond repeatedly to the stimulus to be remembered. In the human literature investigators attribute improved recall with repetition of stimulus material to a rehearsal process and a resulting increase in strength of the memory trace. While it is generally assumed that monkeys are not capable of rehearsal, it is interesting that similar changes in performance in both monkey and man are obtained by arranging conditions so the subject can repeatedly respond to the to-be-remembered stimulus.

THE IMPORTANCE OF STIMULUS DIFFICULTY IN SHORT-TERM RETENTION

In the preceding experiments red and green were used as stimuli in the DMS task. The rather surprising level of retention at even 60-sec delays may be due in part to the nature of the stimuli, and in part to the subjects being so familiar with colors. Certainly, it is well known that monkeys learn to discriminate between colors with relative ease (Warren, 1954). In human STM research level of difficulty and meaningfulness of material have been shown to be important variables (Peterson, Peterson, & Miller, 1961; Pylyshyn, 1965). The more meaningful the material (and presumably the less difficult), the better the retention. The next experiment was designed to evaluate the importance of level of difficulty of stimulus material on STM of monkeys.

For purposes of the present study, level of difficulty of stimulus material was operationally defined by determining the number of trials required to learn to discriminate between pairs of stimuli. Stimuli consisted of colors (red and green), patterns (||| and ≡), and frequency (white lights flashing at 4 and 7 cps). Five experimentally naive stumptail macaques were used as subjects in this experiment. After preliminary training in pressing white lights, the subjects received 100 trials a day with pairs of stimuli to a criterion of 51 correct trials out of 56. The order in which the subjects received the various stimuli differed so that one subject began with colors, and two subjects each started with patterns and frequency. After learning to discriminate between the three sets of stimuli, the subjects learned a discrimination reversal where the previously incorrect stimulus was correct. The results of this discrimination training are shown in Fig. 8. It is apparent that there are real differences between the stimuli in terms of number of trials required to learn the discrimination and therefore in terms of difficulty. Colors were easiest for the subjects with a mean of 146 trials required for discrimination and reversal, while patterns required 367 trials, and frequency was the most difficult with a mean of 925 trials needed to learn the discrimination.

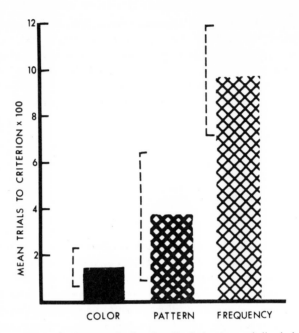

FIG. 8. Mean number of trials to criterion for discrimination and discrimination reversal learning. Standard deviations are indicated by vertical lines.

Since the subjects had more trouble with patterns and frequencies, an attempt was made at each stage of preliminary training in DMS to expose the subjects to these stimuli in an amount proportional to the level of difficulty. The importance of equating level of learning in studies investigating the influence of variables on human STM has been emphasized by Underwood (1964). It soon became apparent that the subjects were having a great deal of trouble with flickering lights and, in fact, were unable to perform a simultaneous match to sample even after 1600 trials. Therefore, the flickering stimuli were dropped from the study. Then DMS was carried out with colors and patterns at randomly presented delays of 0, 1, 5, 12, and 20 sec. A total of 75 trials were given each day with colors and patterns being presented on alternate days.

Results of this experiment are not available since data are still being collected. However, data for the first 14 days of testing show that the percent of the trials on which correct matching responses were made was 73% for colors and 54% for patterns. There was also a suggestion from the data that improvement over days was greater for colors. A more detailed analysis will be carried out; however, it would appear from the preliminary analysis that at this stage of training STM is better for colors than for patterns.

These results suggest that highly discriminable stimuli provide better retention over short intervals of time than stimuli that are not as discriminable. The pair of stimuli comprised of red and green were certainly more discriminable for the monkeys than the patterns and flickering lights. If one assumes that interference between stimuli is the best explanation for forgetting (Keppel & Underwood, 1962), then it seems reasonable to suppose that there would be more interference between stimuli that were less discriminable. Support for such a hypothesis is found in the present study not only in differences in STM for colors versus patterns, but also in the improvement in performance that was found as the animals had more experience with the colors. Presumably, the stimuli would become more discriminable and less difficult with repeated testing.

In a more direct test of the importance of interference in monkey STM, Etkin and D'Amato (1969) varied the sample set size, i.e., the number of stimuli employed as samples (two, three, or four). It was assumed that DMS performance would be poorer with larger sample sets because of proactive interference. However, this hypothesis was not supported since there were no differences in STM between the three sample set sizes and there were no within-days interference effects. Further studies are needed before the importance (or lack of importance) of interference in monkey STM can be determined. Certainly, the monkey is an excellent subject to use in such studies since the experimenter can not only evaluate and control difficulty of the stimuli, but he also has maximum control over previous experience of the subjects with the experimentally presented stimuli.

INTERTRIAL INTERVAL EFFECTS

A variable that is possibly of both theoretical and practical importance is the time between trials. The old question of massed versus distributed practice effects is one of considerable importance in many situations. It has been reported that STM in humans is impaired with 0.5-sec intertrial intervals when compared with retention at 30- and 180-sec intervals (Decker & Allen, 1969). Gleitman, Wilson, Herman, and Rescorla (1963) reported a superiority of distributed practice when rhesus monkeys were tested in delayed response under a massed condition of 20 trials per day and a distributed condition of 2 trials per day for 10 days. However, no difference in performance was found by Fletcher and Davis (1965) when they tested monkeys at 8-, 16-, and 24-sec intertrial intervals. It does seem reasonable that the longer the interval between responses, the less would be the effects of previous trials on subsequent performance. A more practical consideration is the amount of time involved in testing and the possibility that more subjects could be run in a session if intertrial-interval effects are minimal.

In order to determine the effects of various intertrial intervals in our DMS situation, 60 trials a day were given at either 5-, 15-, 30-, or 60- sec intervals. Delays consisted of 0, 5, 15, 30, or 60 sec. The subjects were the same three stumptail macaques used in previous studies.

Results of this experiment are shown in Fig. 9 where percent correct responses are plotted as a function of delays and intertrial intervals. Analysis of these data indicated that better retention is obtained with distributed practice. Specifically, the 5-sec intertrial-interval condition differed from the 15-, 30-, and 60-sec conditions but there were no significant differences between the three longer intertrial intervals. When one compares the magnitude of the difference between massed and distributed practice with the differences in retention found with some of our other experimental manipulations, it is apparent that intertrial interval is an important variable in DMS performance.

CONCLUDING REMARKS

The research summarized above has demonstrated that one can study STM in the monkey by employing experimental procedures similar to those used in research with humans. Further, it is encouraging that the variables studied in these experiments have generally similar effects on STM of both monkey and man. However, a point of some concern is comparability of the DMS task with procedures usually used to study human STM. In addition to the obvious similarities of presentation of the to-be-remembered stimulus, delay, and test for retention, there are some important differences. Because of man's tremendous capacity for storing information, the stimulus material used in the STM experiment is related in many subtle ways to the subject's previous experience. Although the same must apply to monkeys, there should be a real difference in degree. It is usually the case in human STM research that the stimulus material consists of items that are easily verbalized, whereas lower primates are nonarticulate. Another obvious difference is that recall is typically used in human research, while the DMS task as we have used it involves recognition. Many other differences could be listed. It is important to point out that these differences must be taken into consideration in any attempts to quantitatively compare STM for monkey and man.

It is interesting to think about the extent to which monkeys and apes may be better subjects than humans to use if one is really interested in studying STM. Certainly, this would be the case if one wants to investigate the underlying neural bases of STM, but the same may be true for those interested in studying

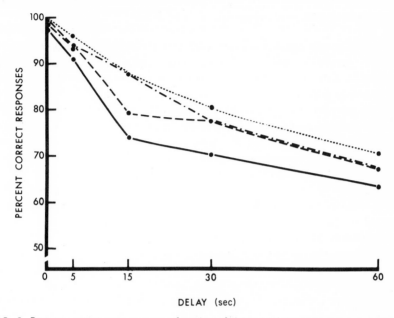

FIG. 9. Percent correct responses as a function of delays and intertrial interval. Intertrial interval: (——) 5 sec; (- - -) 15 sec; (• —•) 30 sec; (• • •) 60 sec.

STM at a behavioral level. One of the real problems encountered in human STM research is control of rehearsal of the material that is to be remembered (Keppel, 1965). It may never be possible in human research to know what kind of interpolated activity will prevent rehearsal and at the same time produce a lack of retroactive interference. Some of the problems involved in defining rehearsal have been discussed by Norman (1969). Although we can not be sure whether nonarticulate primates have a process similar to rehearsal as it exists in people, they would not possess the vocal aspect of rehearsal, and would probably not develop rehearsal strategies that are often reported by human subjects (Norman, 1969). Another advantage of using monkeys and apes is that one can control the stimulus material to which the subjects have been exposed in the past. The interference theory of forgetting states that competing material learned before the experiment or during the retention interval is responsible for forgetting (Adams, 1967). Certainly, we stand a better chance of controlling previous experience with the stimulus material if monkeys and apes are used as subjects.

Besides the theoretical implications of the present research, several of the experimental procedures could be used to advantage in studies designed to see how various brain structures are involved in memory. For example, it is well known that human patients with bilateral lesions of hippocampus have trouble retaining most new information on a long-term basis (Milner, 1968). Although STM appears to be unimpaired, patients forget if they are distracted. Studies with animals have usually not indicated that the hippocampus is involved in memory (Jarrard & Lewis, 1967; Weiskrantz, 1966). However, it would be interesting to look at the effects of different amounts of interfering activity carried out during the delay and different levels of stimulus difficulty on STM of monkeys with lesions of hippocampus. Frontal-lesioned monkeys are impaired on most delayed-response tasks (see Warren & Akert, 1964). The deficit could perhaps be better understood if attempts were made to vary strength of the stimulus trace (by manipulating repetitions of the sample as in our experiment) and/or look at the effects of controlling interval behavior.

The present research has demonstrated that the processes responsible for STM in the monkey can be experimentally manipulated in a way similar to that reported for human subjects. Much research needs to be done before we can decide whether STM in monkey and man differs in principle or only in degree. Answers to questions like the following are needed: Do nonhuman primates possess a rehearsal process similar to the one characteristically described in theorizing about human STM? What is the STM capacity (memory span) of the monkey? What is the role of interference in monkey STM? Are there differences in visual and auditory STM in monkeys? Answers to these and other questions should provide the kind of data that is needed to develop a theory of STM in the monkey. It is hoped that our current efforts will provide the basis for such a theory and a better understanding of how STM varies across species.

REFERENCES

Adams, J. A. *Human memory.* New York: McGraw-Hill, 1967.

Atkinson, R. C., & Shiffrin, R. M. Human memory: a proposed system and its control processes. In K. W. Spence & J. T. Spence (Eds.), *The psychology of learning and motivation: advances in research and theory.* Vol. II. New York: Academic Press, 1968.

Blough, D. S. Delayed matching in the pigeon. *Journal of the Experimental Analysis of Behavior,* 1959, 1, 151-160.

Brown, J. Some tests of the decay theory of immediate memory. *Quarterly Journal of Experimental Psychology,* 1958, 10, 12-21.

Corman, C. D., & Wickens, D. D. Retroactive inhibition in short-term memory. *Journal of Verbal Learning and Verbal Behavior,* 1968, 17, 16-19.

Decker, L. H., & Allen, C. K. The intertrial interval and proactive inhibition in short-term memory. Paper presented at the meeting of the Rocky Mountain Psychological Association, Albuquerque, N. M., May 1969.

Deutsch, J. A. The physiological basis of memory. *Annual Review of Psychology,* 1969, 20, 85-104.

Etkin, M., & D'Amato, M. R. Delayed matching-to-sample and short-term memory in the Capuchin monkey. *Journal of Comparative and Physiological Psychology,* 1969, 69, 544-549.

Fletcher, H. J. The delayed response problem. In A. M. Schrier, H. F. Harlow, & F. Stollnitz (Eds.), *Behavior of nonhuman primates.* New York: Academic Press, 1965.

Fletcher, H. J., & Davis, J. K. Evidence supporting an intertrial interpretation of delayed response performance of monkeys. *Perceptual and Motor Skills,* 1965, **21,** 735-742.

Gleitman, H., Wilson, W. A., Herman, M. M., & Rescorla, R. A. Massing and within-delay position as factors in delayed-response performance. *Journal of Comparative and Physiological Psychology,* 1963, **56,** 445-451.

Harlow, H. F. Primate learning. In Stone, C. P. (Ed.), *Comparative psychology.* (3rd ed.) Englewood Cliffs, N. J.: Prentice-Hall, 1951.

Hebb, D. C. *The organization of behavior.* New York: Wiley, 1949.

Hellyer, S. Frequency of stimulus presentation and short-term decrement in recall. *Journal of Experimental Psychology,* 1962, **64,** 650.

Hunter, W. S. The delayed reaction in animals and children. *Behavior Monographs,* 1913, **2,** 1-86.

Jarrard, L. E., & Lewis, T. C. Effects of hippocampal ablation and intertrial interval on acquisition and extinction in a complex maze. *American Journal of Psychology,* 1967, **80,** 66-72.

Jarrard, L. E., & Moise, S. L. Short-term memory in the stumptail (*M. speciosa*): Effect of physical restraint of behavior on performance. *Learning and Motivation,* 1970, **1,** 267-275.

Jarvik, M. E., Goldfarb, T. L., & Carley, J. L. Influence of interference on delayed matching in monkeys. *Journal of Experimental Psychology,* 1969, **81,** 1-6.

Keppel, G. Problems of method in the study of short-term memory. *Psychological Bulletin,* 1965, **63,** 1-13.

Keppel, G. & Underwood, B. J. Proactive inhibition in short-term retention of single items. *Journal of Verbal Learning and Verbal Behavior,* 1962, **1,** 153-161.

Kulp, R. A. Effects of amount of interpolated activity in short-term memory. *Psychological Reports,* 1967, **21,** 393-399.

Lewis, D. J. Sources of experimental amnesia. *Psychological Review,* 1969, **76,** 461-472.

Malmo, R. B. Interference factors in delayed response in monkeys after removal of frontal lobes. *Journal of Neurophysiology,* 1942, **5,** 295-308.

Melton, A. W. Implications of short-term memory for a general theory of memory. *Journal of Verbal Learning and Verbal Behavior,* 1963, **2,** 1-21.

Meyer, D. R., & Harlow, H. F. Effects of multiple variables on delayed response performance by monkeys. *Journal of Genetic Psychology,* 1952, **81,** 53-61.

Miles, R. C. Delayed-response learning in the marmoset and the macaque. *Journal of Comparative and Physiological Psychology,* 1957, **50,** 352-355.

Milner, B. Preface: Material-specific and generalized memory loss. *Neuropsychologia,* 1968, **6,** 175-179.

Moise, S. L. Short-term retention in *Macaca speciosa* following interpolated activity during delayed matching from sample. *Journal of Comparative and Physiological Psychology,* 1970, **73,** 506-514.

Moise, S. L., & Jarrard, L. E. A computer-controlled system for training and testing primates. *Behavioral Research Methods and Instrumentation,* 1969, **1,** 234-236.

Murdock, B. B. The retention of individual items. *Journal of Experimental Psychology,* 1961, **62,** 618-625.

Murdock, B. B. Distractor and probe techniques in short-term memory. *Canadian Journal of Psychology,* 1967, **21,** 25-36.

Murdock, B. B. Response latencies in short-term memory. *Quarterly Journal of Experimental Psychology,* 1968, **20,** 79-82.

Nissen, H. W., Carpenter, C. R., & Cowles, J. T. Stimulus-versus-response differentiation in delayed reactions of chimpanzees. *Journal of Genetic Psychology*, 1936, **48**, 112-136.

Norman, D. A. *Memory and attention: An introduction to human information processing.* New York: Wiley, 1969.

Peterson, L. R., & Peterson, M. J. Short-term retention of individual verbal items. *Journal of Experimental Psychology*, 1959, **58**, 193-198.

Peterson, L. R., Peterson, M. J., & Miller, A. Short-term retention and meaningfulness. *Canadian Journal of Psychology*, 1961, **15**, 143-147.

Peterson, L. R., Saltzman, D., Hillner, K., & Land, V. Recency and frequency in paired-associate learning. *Journal of Experimental Psychology*, 1962, **63**, 396-403.

Pribram, K. H. Some physical and pharmacological factors affecting delayed response performance of baboons following frontal lobotomy. *Journal of Neurophysiology*, 1950, **13**, 373-382.

Pylyshyn, Z. W. The effect of a brief interpolated task on short-term retention. *Canadian Journal of Psychology*, 1965, **19**, 280-287.

Scheckle, C. L. Self-adjustment of the interval in delayed matching: Limit of delay for the rhesus monkey. *Journal of Comparative and Physiological Psychology*, 1965, **59**, 415-418.

Slamecka, N. J. (Ed.) *Human learning and memory.* New York: Oxford University Press, 1967.

Underwood, B. J. Degree of learning and the measurement of forgetting. *Journal of Verbal Learning and Verbal Behavior*, 1964, **3**, 112-129.

Warren, J. M. Perceptual dominance in discrimination learning by monkeys. *Journal of Comparative and Physiological Psychology*, 1954, **47**, 290-292.

Warren, J. M., & Akert, K. (Eds.) *The frontal granular cortex and behavior.* New York: McGraw-Hill, 1964.

Waugh, N. C., & Norman, D. A. Primary memory. *Psychological Review*, 1965, **72**, 89-104.

Weiskrantz, L. Experimental studies of amnesia. In C. W. M. Whitty and O. L. Zangwell (Eds.), *Amnesia.* New York: Appleton-Century-Crofts, 1966.

Weiskrantz, L., Gross, C. G., & Baltzer, V. The beneficial effects of meprobamate on delayed response performance in the frontal monkey. *Quarterly Journal of Experimental Psychology*, 1965, **17**, 118-124.

Williams, H. L., Beaver, W. S., Spence, M. T., & Rundell, O. H. Digital and kinesthetic memory with interpolated information processing. *Journal of Experimental Psychology*, 1969, **80**, 530-536.

CHAPTER 2

Comparison of Amnesic States
in Monkey and Man

L. Weiskrantz

Certain forms of brain pathology in man lead to extremely striking and persistent forms of anterograde amnesia. Moreover, the amnesia appears to be for material that we would, in a rough sort of way, call "cognitive" in type; it has been claimed, for example, that memory for motor skills is unaffected in such patients, whereas verbal and some perceptual memories are severely affected. Because of the theoretical implications that stem from the alleged properties of the amnesic state, it obviously has been a strong temptation to find similar amnesic states in animals. But the search has been very disappointing. The supposedly critical lesions in animals apparently produce qualitatively different sorts of deficits, and it would be stretching a point to describe these defects as being basically cognitive in character. Having reached this point, most commentators have generally concluded, either with despair or with enthusiasm, that man and monkey are different. But a closer reexamination of the human defect allows one to suggest the animal workers may have been working along the right lines after all, and that their characterization might be applied to the human amnesic state, which might in turn further illuminate the animal work. Because the cognitive capabilities of man and other mammals are different, the defects no doubt have a different appearance, but perhaps they have a common basis.

Confirmed instances of bitemporal damage in man are relatively rare, but the effects are dramatic and have been described frequently (cf. Milner, 1966). The striking aspect of the syndrome is the inability of such patients to remember fresh experiences for more than a few minutes, unless they can sustain them by rehearsal. The incapacity is a severe and persistent one, showing only slight alleviation over a period of 14 years (Milner, Corkin, & Teuber, 1968). The patients have normal "short-term" memory, as judged from their span of attention and their performance on a variety of "two-component" memory tasks designed to yield independent measures of short-term memory (STM) and long-term memory (LTM) (Baddeley & Warrington, 1970). Their memory for preoperative events is said to be normal, except for a variable retrograde amnesia for the few months just preceding surgery. Their intelligence is unaffected, and quite complex problems can be successfully tackled within the limits of their ability to retain information.

So much for the conventional description: it applies not only to bitemporal cases, but also, in its important features, to other instances of the "amnesic syndrome," associated with Korsakoff's psychosis or Wernicke's encephalitis. The conventional description leads one directly and compellingly to a formal description in the following terms: that amnesic patients suffer from an inability to transfer information from STM to LTM, although retrieval from both stores is normal. An alternative formulation is that the patients suffer from a defect of "consolidation" of long-term traces. Although the two alternatives are frequently treated as identical, they need not be. In fact, if the recent report is accepted (Warrington & Shallice, 1969) that there can be an impairment in auditory STM combined with normal LTM, then serial processing of information (whereby it is transferred from short-term to long-term stores) is called into question. But at any rate, the natural description of the syndrome is in terms of a failure of a full-fledged acceptance into long-term storage of new events.

Such a formulation is so dramatic and, in a sense, so surprising that it immediately tempts one to attempt to reproduce it in animals with much better control of lesion site and of past experience. It is dramatic because one knows so little about the physiological foundations of LTM that any lead as to where to concentrate one's efforts is welcome. It is surprising because past efforts to locate the "engram" have merely confirmed its elusiveness; if it is so difficult to locate the engram it is surprising that one can so easily locate the gate that controls entry into it.

A number of efforts have been made to reproduce the syndrome in animals, but before we consider these we must first examine the supposed critical sites of pathology in man. Fortunately, the cases of bitemporal surgery in man are rare, and so unfortunately an answer is difficult to achieve. Penfield's cases (Penfield & Milner, 1958) were, in fact, unilateral temporal lobectomies associated, it is thought, with pathology of the other temporal lobe, indirect evidence in support of which are the temporary amnesic effects of anesthetization by sodium amytal of one hemisphere in different patients already having known temporal pathology in the contralateral hemisphere (Milner, Branch, & Rasmussen, 1962). But the unilateral temporal lobectomies were large removals that included lateral and ventral neocortex as well as hippocampal tissue medially. The suggestion that the medial portions are critical comes from the surgical approach of Scoville, whose lesions were deliberately placed medially (Scoville & Milner, 1957). From his drawings the lesions included hippocampus and also some neocortical tissue on the ventro-medial margin of the lobe, and undoubtedly a certain amount of fiber damage to underlying white matter.

The hippocampus also gains importance from the studies of unilateral temporal damage by Milner and her colleagues. Unilateral lesions (in the absence, it must be assumed, of contralateral pathology) do not yield the strikingly dramatic results associated with bilateral pathology, but memory impairments appear to be one of their main concomitants, the type depending on whether the left dominant or right subdominant hemisphere is damaged. Right temporal damage is associated with visual, auditory, and tactile memory impairment, left temporal with verbal memory impairment (Milner, 1967; Warrington & James, 1967). The right temporal syndrome does not seem to be accounted for by perceptual impairments per se, which are more parsimoniously associated with posterior temporal and parietal pathology (Warrington & Rabin, 1970). Similarly, left temporal damage need not produce a failure of word recognition as such in order to yield a verbal memory impairment. The important point is that in both cases it has been claimed that hippocampal damage accentuates the behavioral impairment, and in the right temporal case disproportionately so in comparison with nonhippocampal damage (Milner, 1968). We do not know whether the bilateral temporal syndrome can be assumed to be a simple composite of the effects of the two unilateral temporal syndromes, the bilateral effects appearing to be much more dramatic because of the inability of the subject to use one normal memory system to compensate for the impaired one (e.g., by using verbal labels for impaired perceptual memory), but that is the most parsimonious assumption.

The clinical evidence, therefore, suggests a special importance for the hippocampus, and this would seem to be the first structure to try to involve lesion studies in animals in attempting to replicate the human bitemporal syndrome. But it should be stressed that our evidence on localization is very far from satisfactory. No lesion restricted to hippocampus alone can be expected from any mortal surgeon and, generally speaking, the more complete the hippocampal removal the greater the probability of damaging surrounding tissue; nor is "size" of lesion very easy to assess meaningfully in this region of the brain—a small undercut might disconnect a much greater mass of tissue than immediately overlies the cut, and in fact, quite independently of size, a disconnection might be more devastating than removal of tissue as such. It is precisely because just such questions cannot be answered decisively by the clinical cases that one tries to do more controlled experiments using animals.

But, alas, so far as effects on memory are concerned, hippocampal lesions in animals have been woefully disappointing. A variety of descriptions of the effects of such lesions in animals has been suggested, but few workers have been rash enough to conclude that an inability to establish long-term traces is one of them. A few studies have explicitly attempted to test monkeys under conditions thought likely to replicate the human bitemporal defect. For example, Orbach, Milner, and Rasmussen (1960) found no defect in medial bitemporal monkeys on a discrimination learning task in which widely spaced trials were filled with massed trials in irrelevant discriminations. Nor could they find more than a mild impairment on the retention of postoperatively acquired tasks. Kimble and Pribram (1963) found no impairment in bilateral hippocampectomized monkeys on discrimination learning tasks with intertrial intervals of 6 min. Indeed, hippocampectomized rats learn certain tasks even faster than controls, e.g., two-way active avoidance tasks (Isaacson, Douglas, & Moore, 1961). A number of impairments in learning and retention have been reported to be associated with hippocampal damage in animals but these can be accounted for largely in two ways: first, as a by-product of nonmnemonic changes that are positively associated with such lesions; and second, as a by-product of unavoidable damage to neighboring neocortex and the pathways connected with it.

So far as the positive effects of hippocampal damage are concerned, there is now a fair measure of agreement among workers on rats and monkeys. Such operates are slower to extinguish responses in a variety of discrimination, runway, and other operant situations. They are slower to acquire discrimination reversals. They tend to be indistractible by novel stimuli under some conditions, and show a diminished level of spontaneous alternation and a deficit in acquiring learned alternations. They are also deficient in passive avoidance situations. Not all these results have been obtained on both species, but there is a good measure of agreement in the literature. (On the other hand, one must guard against any easy assumptions of uniformity among mammals: Jarrard and Bunnell (1968) have shown that hippocampal lesions in the hamster are different from those in rats on their effects on open-field behavior.) Reference to the relevant studies for the rat can be found in Kimble (1969) and in Douglas and Pribram (1966) for the monkey.

Various unitary interpretations have been made of this cluster of effects. Thus, Kimble (1969) suggests that "the mammalian hippocampus [constitutes] part of the neural machinery necessary for the generation of a brain process which is functionally equivalent to Pavlovian internal inhibition." Douglas and Pribram (1966) postulate that the hippocampus acts on a process so as "to diminish awareness of experience as a function of the probability of nonreinforcement." Most accounts have something of the same flavor, and agree that hippocampectomized animals will be less likely than controls to stop doing something that is either punishing or nonrewarding, and more likely to continue to do something that is or has been rewarding. Whether the perseveration is best thought of as occurring on the response side or on the attentional side of the equation is still unsettled, but it is tempting for many reasons to subscribe to Crowne and Riddell's (1969) recent conclusion that "damage to the hippocampus impairs the ability to orient or shift attention to a new stimulus when S is engaged in the performance of some powerful ongoing response" (p. 748).

But, whatever the interpretation, such an impairment is certainly not an inability to hold information in long-term storage; if anything, it is an inability to eliminate information. It is true that, depending on the fine details of the situation, it is of course a defect that can lead to impaired learning or retention, as in the case of passive avoidance, but the mere fact that learning can be impaired does not make the animal hippocampal defect relevant to the human amnesic syndrome—learning situations can reflect practically anything that is wrong with an animal. It is the lack of apparent connection between the human and animal hippocampal impairments that has been our continuing dilemma. Because of this dilemma, some years ago we considered the second possibility that I referred to above: that the amnesic syndrome in man was caused by inadvertent damage to neighboring neocortex or its connections. Here, at least for the monkey, we did have a promising candidate in the inferotemporal cortex, and the homolog of such cortex was almost certain to have been involved in the Penfield and Scoville surgical cases. On the other hand, it is relatively easy to study inferotemporal lesions in the monkey without damaging the hippocampus.

Inferotemporal lesions were of interest to us because of the well-established impairment in visual discrimination learning associated with them (Mishkin & Pribram, 1954). While it is impossible to prove the universal negative, such sensory capacities as have been measured in these animals are unchanged (e.g., Ettlinger, 1959; Weiskrantz & Cowey, 1963; Symmes, 1965; Cowey & Weiskrantz, 1967). In addition, the learning impairment seemed qualitatively different from those caused by hippocampal defects (e.g., delayed alternation is normal) and Mishkin (1954) had shown that the inferotemporal learning impairment was much more severe than that caused by hippocampal lesions with minimal neocortical damage. Everything pointed to a defect that was primarily of visual learning. The question we asked was whether that learning impairment could be said to be the result of a failure of LTM. The relevant experiments have been published and need no lengthy review here (Weiskrantz, 1967; Iversen & Weiskrantz, 1964, 1970). The general requirement was to design a paradigm that would allow rapid "short-term" learning of many problems within a few trials, with the possibility of studying their retention at various intervals. There were both inferotemporal operates and operates with inferotemporal plus hippocampal damage. The results were quite clear in demonstrating deficient savings after an interval of 24 hr in the temporal lobe operates, and these poor savings could not be accounted for in terms of an initial learning impairment. Even after an interval of 15 min savings were poor.

This, to say the least, gave us grounds for thinking we were making some progress in relating the animal and human syndrome. Meanwhile, independent work had shown that the inferotemporal defect is divisible into an anterior and posterior defect. Iwai and Mishkin (1967) and Cowey and Gross (1970) have shown that anterior lesions produce impairments on "concurrent" (i.e., serial) discrimination learning and difficult color discrimination learning, whereas posterior lesions result in impaired learning of pattern discriminations. The defect in

serial learning, in which the animals were trained on several object discriminations within each testing session, is interpreted by Mishkin and Iwai as one in
"memory or associative learning." Interestingly, the same defect had been reported by Correll and Scoville (1965) to result from medial temporal lesions in
monkeys (hippocampus plus neighboring structures including neocortex)
modeled on those imposed by Scoville on human patients. Whether anterior
inferotemporal lesion effects can be dissociated from those of hippocampal
lesions in the monkey we do not know as yet.

But in considering our own results together with the serial learning results,
there still seems to be a large apparent gap between monkey and man. First, the
monkey memory defect is modality-specific so far as one can tell—at least Iversen (1967) showed the savings for tactile problems was not impaired using the
same paradigm that produced poor savings for visual problems. This is, in itself,
perhaps not too worrying so long as one can at least demonstrate formal similarities between man and monkey in one modality as far as LTM is concerned.
There might be a number of reasons why the human defect might be multi-
modal, which need not be discussed here. But a second line of evidence appears
to show that even within the visual mode there are qualitative dissimilarities
between monkey and man. The paradigm originally used by Dr. Iversen and
myself was very convenient for testing lots of problems thereby building up a
statistical picture rapidly, but it contained within it the possibility of massive
retroactive and proactive interference effects. When just a single problem was
given in a session, with intervening darkness for 15 min before testing for retention, the operated monkey's retention was virtually normal. In contrast, when
another visual problem was presented in the interval, savings were impaired. The
importance of interference was also revealed in Correll and Scoville's (1965)
study of serial discrimination learning in medial temporal lobe monkey operates.
Their control animals required the same or even fewer trials to learn each additional problem beyond the first problem, whereas the lesioned animals required
more trials for additional problems. Negative results might also be taken to
support the same thesis: Butler (1969) presented single problems in each session
to inferotemporal operates (in a split-brain preparation) and found normal
savings after 24 hr. Clearly we cannot be dealing with a failure of retention in
long-term storage if the memory defect is only demonstrable when there is
interference by other items that have gained entry into long-term storage themselves. Milner, indeed, has argued against the view that the memory impairment
in man does not depend critically upon distractions in the delay period (e.g., cf.
Milner, 1968).

A third and related line of evidence also disturbed us. If visual information in
the operates were mainly available only in a short-term store, then there ought
to be abnormally rapid failure of retention once a certain shortish interval is
exceeded. In order to look at this, forgetting curves were measured in monkeys

with temporal lobe lesions similar to those used by Iversen and myself. The monkeys were taught a simple object discrimination, then kept in darkness for varying intervals ranging from 0 sec up to just over 2 hr. This was done for 50 problems. When we analyzed the results in the conventional way by plotting savings in terms of numbers of trials required for relearning relative to original learning trials, we did indeed find more rapid forgetting in the operated animals (see Fig. 1). In fact, on one method of calculation their savings were zero within

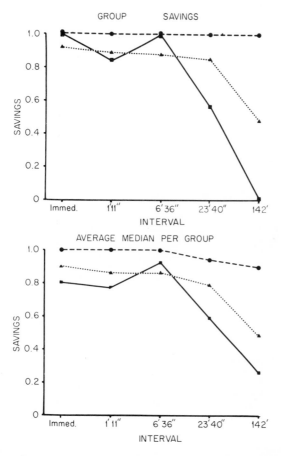

FIG. 1. Savings as a function of interval following learning to criterion. Score calculated by (learning trials − relearning trials)/ (learning trials + relearning trials). Top graph: median score for all group scores combined. Bottom graph: average of median scores for individual animals in each group. Ten problems per animal were tested at each interval. Dashed line: normal control group. Dotted line: inferotemporal lesion group. Solid line: inferotemporal plus hippocampal lesion group.

approximately 2 hr. Although admittedly the time constants were far outside what would be expected from human work on verbal STM, the results at least went in the right direction for supporting a hypothesis of faulty LTM. But when we considered the results in greater detail, we realized that the conventional savings measure as used typically in animal work is an impure and possibly quite misleading measure (Weiskrantz, 1968b). An animal can have poor savings not only because it forgets more than controls and therefore has more to relearn, but alternatively because it does not forget any more than a control but is slower to relearn from the initial retention level. It is clear that one must plot the relearning curves and examine not only their initial starting points, which give the greatest information for the analysis of amnesia, but their rate of growth as well.

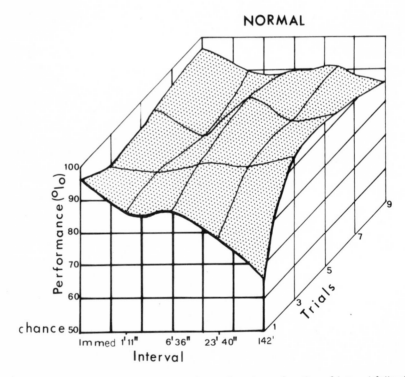

FIG. 2. Relearning curves of normal monkeys plotted as a function of interval following original learning to criterion, including immediate relearning after no interval. Performance scores during relearning (y axis) are plotted separately for odd-numbered trials from 1 to 9 (z axis). Intervals between achievement of criterion in original learning and beginning of relearning are shown in x axis. Intervals in this and Fig. 1 were originally chosen to be equal log units, having values of 0, 1, 5, 25 min, and 2 hr 5 min (except for first interval for which a log value is indeterminate), but deviated in practice from this because of a miscalibration of a multirange commercial timer. Corrected calibrated times are shown in figure. Scores of three monkeys on 50 problems each are averaged. (After Weiskrantz, 1968b.)

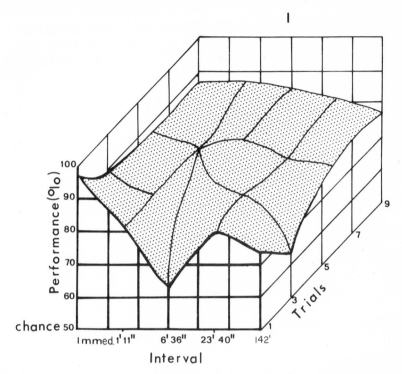

FIG. 3. Same as Fig. 2, except that monkeys had inferotemporal cortex lesions. (After Weiskrantz, 1968b.)

In Figs. 2, 3, and 4 the relearning performance surfaces are plotted as a function of learning-retention interval and relearning trials.

One point is instantly clear: although all groups reached the same criterion of 18 correct out of 20 in learning, the I + H (inferotemporal + hippocampal-lesioned) animals did not show 90% performance even immediately afterward. Why this should be is not obvious, although it no doubt was influenced by their original learning having been slower, because in such a case a chance deviation upward to criterion level is likely to be followed by a drop below criterion level. But the question that concerns us immediately is whether there is or is not faster forgetting given that the I + H group's curve is lower to start with, i.e., in one sense, they have less to forget. Figure 5a shows the forgetting curves based on trial-1 performance only, and in Fig. 5b the I + H curve has been shifted so that its starting point is matched with the control curve. It is possible that with more than 10 problems per point per animal (i.e., a total 30 problems per point) a significant difference would emerge, but it is difficult to have any confidence in

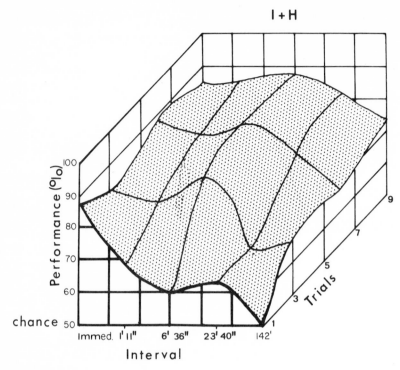

FIG. 4. Same as Fig. 2, except that monkeys had inferotemporal cortex plus hippocampus lesions. (After Weiskrantz, 1968b.)

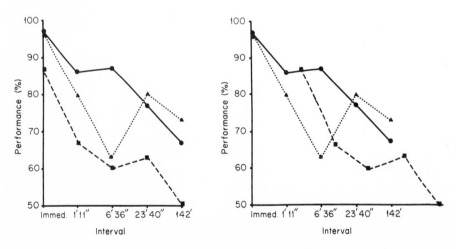

FIG. 5. (*Left*): Trial-1 performance in relearning as a function of interval following initial learning. Same animals as for Figs. 2, 3, and 4. (*Right*): Same, but starting point for group I + H shifted to match normal control value. (●)N; (▲)I; (■)I + H.

the three curves displaying anything other than random fluctuation from a common population.[1]

At this stage, to put it no more strongly, we felt a certain lack of enthusiasm regarding the likelihood of demonstrating a parallel between bitemporal lesions in man and monkey. But we could not quite overlook the fact that the one consistent finding with inferotemporal monkeys, after all, is that they have a difficulty in visual learning, and difficulty in learning is one indisputable failing of the human amnesic: their most striking feature is shared. Perhaps we had been accepting modern fashion too readily in our animal experiments in adopting the current mode of interpretation of memory phenomena: that there is a sequential transfer from STM to LTM. I have already cited evidence that forces one to question such a view (Warrington & Shallice, 1969) and, moreover, the evidence for a process in animals strictly comparable to that commonly labelled STM in man can be questioned (Weiskrantz, 1970). Perhaps it is still the case that both monkeys and men with neocortical temporal lobe damage suffer from a relative difficulty in getting new information into long-term storage, but that what gets in is stored normally and without precipitous decay. It may only be because man alone has an emergency verbal rehearsal system, with its own peculiar limited capacity, that one sees the appearance of rapid forgetting in the human temporal lobe case whenever there is inadequate learning. Therefore, if one could somewhat succeed in getting some information into the amnesic that does not depend critically on such a rehearsal process for its survival, perhaps he will show a syndrome comparable to the inferotemporal monkey: that is, slower learning but good retention of what is learned.

Thus Dr. Elizabeth Warrington and I set about considering the forgetting curves of human amnesics. What we wanted to see was whether, if their performances were matched at the end of the learning phase, their forgetting curves would be seen to be no more rapid than controls. Our patients were mostly cases of Korsakoff's psychosis but one of them was a temporal lobe operate. We hoped that by giving verbal material with varying amounts of repetition we would be able to achieve a sufficient degree of learning to measure their forgetting curves over varying intervals, up to 15 min. In the main aim of the experiment we failed: we were not able to secure high enough levels of learning in the amnesics, even with many repeated trials, to be able to match them against controls given only a few trials, and therefore we could not compare the forgetting curves of the two groups with any confidence (Warrington & Weiskrantz, 1968b).

---

[1] The bumps on the forgetting curves do have some possible interest from the point of view of consolidation theory, and I am grateful to Dr. J. McGaugh for drawing my attention to the matter. The bumps are not statistically significant but it is possible to speculate that they might reflect the build-up of a long-term consolidation process which is retarded in the temporal operates.

In the end, as you shall see, we did eventually succeed but only by changing our tack altogether. But in this first experiment, one interesting point did emerge in the verbal recall data: we found a large number of intrusion errors, and of those intrusion errors 50% were words from prior lists, seen in some cases the day before and sometimes several days before. From persons who apparently cannot remember beyond a few minutes, this was indeed a surprise. Such intrusions from LTM in amnesic patients have also been found more recently by Baddeley and Warrington (1970) and Starr and Phillips (1970). Evidently something must be getting stored by the patients for it to show up as an error later.

Because it was so dreadfully tedious trying to teach the amnesics with conventional verbal learning methods, we followed up a suggestion by Dr. Warrington about a method that might be less tedious. Not only did the method work in that sense but it has unexpectedly opened up a large number of new avenues to us and puts the syndrome in a different light. The method (Warrington & Weiskrantz, 1968a) makes use of fragmented drawings of the type used by Gollon (1960) for a perceptual test, and is reminiscent of Williams' use of pictorial "prompts" with patients (Williams, 1953). The subject is required to identify a particular drawing with incomplete information. With repeated trials, however, it was found that subjects required progressively less information to identify items until even the most fragmented forms sufficed. One can measure the rate of improvement to study learning and its retention over time to study memory. The improvement we found is not just a general practice effect; it was quite specific to the items to which the subjects were exposed. The test is very easy to administer and relatively free from stress, especially for amnesic patients, because it is not even necessarily recognized as a memory test but just as a sort of guessing game.

We used both fragmented pictures and fragmented words, as shown in Fig. 6. To our surprise but pleasure, there was clear evidence of learning in the patients for both kinds of material, as well as clear evidence of savings after intervals of 24 hr (Figs. 7 and 8). It is true that the patients learned more slowly for the fixed number of trials given, but from the data obtained in that experiment it is impossible to say whether they would forget more than controls over 24 hr if the performance levels were matched. In a later experiment (Weiskrantz & Warrington, 1970a) using only the fragmented words, we trained amnesics and control patients to the same criterion of two errorless runs, and had good evidence that on immediate retention both groups were at criterion. We were then able to measure and compare forgetting curves more directly for the two groups, that is, to do the experiment we first wanted to do with conventional methods. We did find (Fig. 9) less savings in patients, but the impressive result was that savings were seen even over 72 hr. In fact, at that point there was no significant difference between controls and amnesics. In more informal testing we have found evidence in our temporal lobe case (whose memory impairment was first described clinically by Dimsdale, Logue, and Piercy, 1964) for retention over

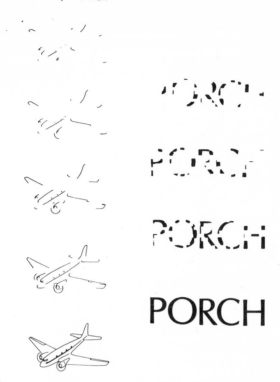

FIG. 6. Fragmented drawings of pictures and words. (From Warrington and Weiskrantz, 1968a.)

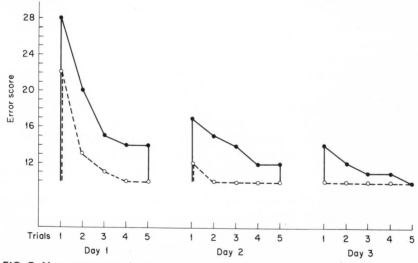

FIG. 7. Mean error scores for each trial for each patient group on the fragmented picture test. Solid line: amnesics. Dotted line: controls. (From Warrington and Weiskrantz, 1968a.)

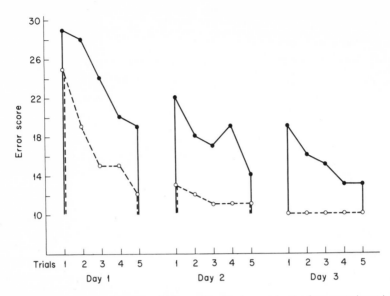

FIG. 8. Mean error scores for each trial for each patient group on the fragmented word test. Solid line: amnesics. Dotted line: controls. (From Warrington and Weiskrantz, 1968a.)

several weeks and even months, although typically the patient denies having done the test before or even recognizing the experimenter. It is important to stress that the same words presented in conventional memory drum fashion yielded only minimal learning and virtually no retention within minutes.

Milner's well-known case, H. M., on whom much of the detailed testing of the syndrome has been based heretofore, was later tested briefly on the fragmented drawings method by Milner et al. (1968) and he was also shown to remember over 1 hr (the only interval used). Milner (1968) speculated that perhaps not only motor skill learning but also "perceptual" learning was spared by hippo-campal damage. In fact, motor learning does not appear to be normal in H. M. (Corkin, 1968) but at least positive savings have been demonstrated (Milner et al., 1968). But to examine the suggestion that the method was one of "percep-tual learning," we constructed a different form of the test that did not place any perceptual strain at all on the patients. Instead of using fragmented letters, we used clear whole letters but presented only the first two or three letters of a five-letter word. Dr. Warrington and I have demonstrated the same pattern of learning and retention phenomena using a 1 hr interval (Fig. 10). The difference between this method and others is not in its content, in whether it is, for example, primarily verbal or perceptual, but in its use of partial information (which can itself be verbal) as a cue for the identification of the whole. Un-doubtedly there are a number of extensions to other forms of content as well.

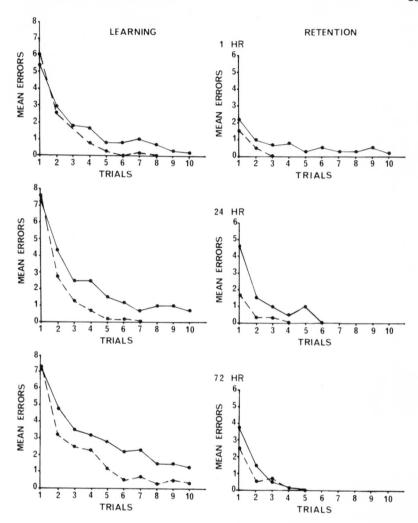

FIG. 9. Mean error scores for each trial during learning and relearning, using fragmented words, for three different intervals following learning. All subjects learned to a criterion of two errorless trials, but only the first 10 learning trials are plotted. (———) amnesics; (- - -) controls. (From Weiskrantz and Warrington, 1970a.)

But taking the results so far, a number of issues are forced. First, perhaps there is nothing surprising about the results at all. What we have succeeded in doing, it could be argued, is to have hit on a particularly sensitive method of measuring memory in patients who need not be considered to have an absolute defect. But, all the same, the argument would continue, their memory is defective as witnessed by the slower learning and poorer savings. This was essentially the conclusion that Williams came to in using a method of "progressive prompts" with amnesic patients—that the patients were still relatively just as impaired even though in absolute terms both patients and controls could be helped to retrieve more with the prompting method. Differences in test sensitivity are well known for their ability to mislead neuropsychologists (cf. Weiskrantz, 1968a); and perhaps we are providing yet another example.

A more difficult issue is whether the dramatic improvement in amnesic patients using the partial information method stemmed mainly from the form of the material at the learning stage or at the retention stage. That is, did patients remember because the information was originally presented in a particular form, or because it was available to them in that form at the retrieval stage, or both? The experiment designed to provide an answer to the second question also, as it happens, answered the first question about test sensitivity. In a balanced pair of experiments (Warrington and Weiskrantz, 1970) the modes of presentation at the learning and retention stages were varied independently. In the first experiment words were presented for learning in the fragmented form shown in Fig. 6. After 1 min (during which the subject counted backward) material for retention was presented in the same fragmented form, by free recall, or by conventional

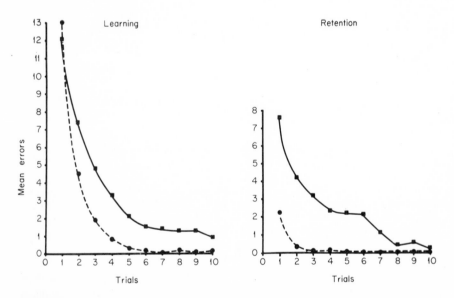

FIG. 10. Mean error scores for each trial during learning and relearning using "letter-by-letter" method. Relearning was after a 1 hr interval. All subjects learned to a criterion of two errorless trials, but only the first 10 learning trials are plotted. (●) Controls; (■) amnesics. (From Weiskrantz and Warrington, 1970b.)

recognition testing of the whole words in a longer list that included some words not shown in the learning series. The results were that the patients were significantly poorer under recall and recognition conditions and not impaired under the fragmented-form condition. In the second experiment the learning phase consisted of lists of whole words shown to the subjects who read them aloud, each list being repeated three times. There were four retention methods, the conventional ones—recall and recongition—and the two methods of partial information—fragmented letters or whole letters forming parts of words. The results in the second experiment were striking. Again the conventional methods yielded significantly poorer scores for the amnesics than for controls; neither of the two partial information methods yielded a difference between groups. Moreover, the pattern of results for patients and controls were quite different. Of all the retention conditions, controls were best on the conventional recognition test, whereas patients were best on the "letter-by-letter" method. The interaction between groups and retention methods in analysis of variance was significant, providing formal evidence in favor of a differential reaction of the groups to the same test material. Thus, the partial information method does not succeed merely by being a more sensitive test. The important positive point is that the form of the material at the learning stage is irrelevant: all that matters for adequate retention is that it be in a particular form at the retention stage itself. Indeed, as we have seen, the difference between groups disappeared in the last two experiments when partial information methods were used at the retention stage.

The second experiment also disposes of another interpretation, and at the same time exposes a striking paradox. Because amnesic subjects were so poor on the conventional recognition method (in fact, earlier we found suggestive evidence that patients were relatively poorer on recognition than on recall; Warrington & Weiskrantz, 1968b), their adequate performance under more favorable conditions can hardly be a removal of a blockage in finding the names as such; that is, we are not dealing with a "tip of the tongue" phenomenon. Paradoxically, in fact, patients, having received whole words during learning, do better in retrieving them when given partial information about them than when they are given the whole word in a recognition task!

What can one make of all this? What does the partial information method have that other methods do not? The critical difference that impresses Dr. Warrington and me is this: the partial information may be partial but it is relatively unambiguous in the sense that, within limits, it either does or does not match the remembered item to be identified. The patient thereby has a good method of eliminating incorrect answers before making them. A particular fragmented picture matches a remembered airplane much better than it matches anything else, and a patient is unlikely to say, for example, "elephant" when being shown the picture. But we have seen that in a conventional free recall situation he is indeed likely to say elephant, particularly if it appeared in a previous list. Conventional recognition tasks are especially difficult in this respect—all the items are familiar, including the ones that were not shown just previously. In fact, the degree of certainty that partial information offers in eliminating false responses can be put under control by varying the information content in the "letter-by-letter" situation, and there is no reason why in principle they could not be eliminated completely.

Why should amnesic patients be helped by having a method of eliminating false positives? There are three possibilities, which we just have time to mention. The first is that the information in store is misclassified in broad categories so that the wrong item emerges because it shares a file with the correct item. The work of Butter, Mishkin, and Rosvold (1965), for example, showing broader generalization gradients in inferotemporal monkey operates, would be consistent with this view. This is not an easy hypothesis to test in man, but Dr. Warrington and I did investigate a variety of categorization procedures in our amnesic patients along the lines, among others, of the work suggested by Tulving and Pearlstone (1966) and Mandler and Pearlstone (1966). Our evidence produced no support for the hypothesis that the patients misclassified. They had poorer retention, but it was just as sensitive to the same variables as controls. The second and third possibilities are related. Both assume that potentially incorrect responses have excess strength in amnesics. On one view the traces themselves are stronger because of lack of dissipation, i.e., lack of forgetting. On the other view, the traces are of normal strength but the responses are disinhibited. We do not have any evidence as yet to favor one view or the other, although we think the two hypotheses are experimentally discriminable. The generalization results are consistent with either of these views (and with others as well), as are some interesting incidental observations of Iwai and Mishkin (1967) in reporting the dissociation between anterior and posterior inferotemporal operates. When they carried out equivalence testing for different portions of the pattern discrimination, they found that controls tended to extinguish their preferences during equivalence training, no doubt because both stimuli were rewarded. The anterior temporal operates continued to discriminate the previously positive cue, and failed to extinguish their preference. Douglas and Pribram (1966) also found slower extinction of a learned discrimination by hippocampal operates; similarly, Kimble and Kimble (1970) found greater persistence of "hypothesis" behavior in hippocampectomized rats.

You will see that we have come more than a full circle. We started with an assumed human defect of long-term storage for which the hippocampus was implicated, and saw that the animal literature at best supported quite another view of the effects of hippocampal damage. Nor could the effects of neocortical damage easily be interpreted as an abnormally rapid decay in LTM, although poor retention was seen with an interference paradigm. But the monkey with anterior neocortical temporal lobe damage does consistently display poor visual learning, and especially so when there is full scope for interference in the acquisition phase. Interference effects in the learning phase are also disruptive in animals with medial temporal lesions, which include both hippocampus and direct or indirect damage to neocortex. We do not know as yet whether anterior inferotemporal effects can be dissociated from the effects of hippocampal lesions in the monkey. If they can be, we can speculate that the medial temporal syndrome in the monkey will emerge as a composite of at least two disorders, the one a disinhibition of established information in memory, the other an impairment in the processing and establishment of new information.

When we turn to the human amnesic patient, however, we find that under certain circumstances he is far from amnesic. Moreover, the most parsimonious assumption about how his retention can be helped by those special circumstances suggests that he is suffering from a condition qualitatively very similar to that produced in animals by hippocampal damage. Whether there is an additional neocortical component cannot be assessed at the present time. No doubt the cognitive capacities of man and monkey are different, but in both cases their respective capacities appear to be disinhibited or rendered unduly persistent by medial bitemporal lobe damage. With man's elaborate linguistic skills, such an effect can be disastrous for everyday memory because of the acknowledged importance of interference phenomena in human verbal memory.

There is just one further point to make. Slowly a catalog of exceptions to absolute amnesia in amnesic patients is being assembled. Motor learning skills (e.g., rotary pursuit, tracking, etc.) can be learned and retained by H. M., if not altogether normally. Very simple tactile and visual mazes can also be learned, but with a severe deficit. And now we see that verbal learning is possible with the methods of partial information, with only a slight defect and sometimes none at all. Perhaps we will never have anything better than a catalog. But it is interesting to speculate about the possibility that the exceptions may share common properties: either a minimal degree of within-task interference, or some method of eliminating potentially incorrect responses. It is worth noting that even in the simple maze H. M.'s performance at the first few choice points was good; errors built up at the later choice points (Milner et al., 1968). Motor learning tasks such as tracking may well have their own correction devices built into them equivalent to the one that we have postulated for the partial information methods.

It is to some degree a matter of fashion and preference whether one emphasizes the continuities or the discontinuities between animals and men. At the moment the discontinuities appear to be enjoying a certain vogue. I hope that this paper will at least indicate how some of the undoubtedly genuine built-in differences can be shown to be sensitive to similar forces, without at the same time forcing us to conclude that man is a monkey.

Acknowledgments

The author would like to express his appreciation to Drs. A. Cowey, D. P. Kimble, and Elizabeth Warrington for their helpful comments.

REFERENCES

Baddeley, A. D. & Warrington, E. K. Amnesia and the distinction between long and short-term memory. *Journal of Verbal Learning and Verbal Behavior*, 1970, 9, 176-189.

Butler, C. R. Is there a memory impairment in monkeys after inferior temporal lesions? *Brain Research*, 1969, 13, 383-396.

Butter, C. M., Mishkin, M., & Rosvold, H. E. Stimulus generalization in monkeys with inferotemporal and lateral occipital lesions. In D. J. Mostofsky (Ed.), *Stimulus generalization*. Stanford, Calif.; Stanford University Press, 1965.

Corkin, S. Acquisition of motor skill after bilateral medial temporal excision. *Neuropsychologia*, 1968, 6, 255-265.

Correll, R. E., & Scoville, W. B. Effects of medial temporal lesions on visual discrimination performance. *Journal of Comparative and Physiological Psychology*, 1965, 60, 175-181.

Cowey, A., & Gross, C. G. Effects of foveal prestriate and inferotemporal lesions on visual discrimination by rhesus monkeys. *Experimental Brain Research*, 1970, 11, 128-144.

Cowey, A., & Weiskrantz, L. A comparison of the effects of inferotemporal and striate cortex lesions on the visual behaviour of rhesus monkeys. *Quarterly Journal of Experimental Psychology*, 1967, 19, 246-253.

Crowne, D. P., & Riddell, W. I. Hippocampal lesions and the cardiac component of the orienting response in the rat. *Journal of Comparative and Physiological Psychology*, 1969, 69, 748-755.

Dimsdale, H., Logue, V., & Piercy, M. A case of persisting impairment of recent memory following right temporal lobectomy. *Neuropsychologia*, 1964, 1, 287-298.

Douglas, R. J., & Pribram, K. H. Learning and limbic lesions. *Neuropsychologia*, 1966, 4, 197-220.

Ettlinger, G. Visual discrimination with a single manipulandum following temporal ablations in the monkey. *Quarterly Journal of Experimental Psychology*, 1959, 11, 164-174.

Gollon, E. S. Developmental studies of visual recognition of incomplete objects. *Perceptual and Motor Skills*, 1960, 11, 289-298.

Isaacson, R. L., Douglas, R. J., & Moore, R. T. The effect of radical hippocampal ablation on acquisition of avoidance response. *Journal of Comparative and Physiological Psychology*, 1961, 54, 625-628.

Iversen, S. D. Tactile learning and memory in baboons after temporal and frontal lesions. *Experimental Neurology*, 1967, 18, 228-238.

Iversen, S. D., & Weiskrantz, L. Temporal lobe lesions and memory in the monkey. *Nature*, 1964, 201, 740-742.

Iversen, S. D., & Weiskrantz, L. An investigation of a possible memory defect produced by inferotemporal lesions in the baboon. *Neuropsychologia*, 1970, 8, 21-36.

Iwai, E., & Mishkin, M. Two inferotemporal foci for visual functions. Paper presented at the annual meeting of the American Psychological Association, Washington, D.C., 1967.

Jarrard, L. E., & Bunnell, B. N. Open-field behavior of hippocampal-lesioned rats and hamsters. *Journal of Comparative and Physiological Psychology*, 1968, 66, 500-502.

Kimble, D. P. Possible inhibitory functions of the hippocampus. *Neuropsychologia*, 1969, 7, 235-244.

Kimble, D. P., & Kimble, R. J. The effect of hippocampal lesions on extinction and 'hypothesis' behavior in rats. *Physiology and Behavior*, 1970, 5, 735-738.

Kimble, D. P., & Pribram, K. H. Hippocampectomy and behavior sequences. *Science*, 1963, 139, 824-825.

Mandler, G., & Pearlstone, Z. Free and constrained concept learning and subsequent recall. *Journal of Verbal Learning and Verbal Behavior*, 1966, 5, 126-131.

Milner, B. Amnesia following operation on the temporal lobe. In C.W.M. Whitty & O. L. Zangwill (Eds.), Amnesia. London: Butterworth, 1966.

Milner, B. Brain mechanisms suggested by studies of temporal lobes. In F. L. Darley (Ed.), Brain mechanisms underlying speech and language. New York: Grune & Stratton, 1967.

Milner, B. Visual recognition and recall after right temporal-lobe excision in man. Neuropsychologia, 1968, 6, 191-209.

Milner, B., Branch, C., and Rasmussen, T. Study of short-term memory after intracarotid injection of sodium amytal. Transactions of the American Neurological Association, 1962, 87, 224-226.

Milner, B., Corkin, S., & Teuber, H. -L. Further analysis of the hippocampal amnesic syndrome: 14-year follow-up study of H. M. Neuropsychologia, 1968, 6, 215-234.

Mishkin, M. Visual discrimination performance following partial ablations of the temporal lobe: II. Ventral surfaces vs. hippocampus. Journal of Comparative and Physiological Psychology, 1954, 47, 187-193.

Mishkin, M., & Pribram, K. H. Visual discrimination performance following partial ablations of the temporal lobe: I. Ventral vs. lateral. Journal of Comparative and Physiological Psychology, 1954, 47, 14-20.

Orbach, J., Milner, B., & Rasmussen, T. Learning and retention in monkeys after amygdala-hippocampus resection. Archives of Neurology, 1960, 3, 230-251.

Penfield, W., & Milner, B. Memory deficit produced by bilateral lesions in the hippocampal zone. Archives of Neurology and Psychiatry, 1958, 79, 475-497.

Scoville, W. B., & Milner, B. Loss of immediate memory after bilateral hippocampal lesions. Journal of Neurology, Neurosurgery and Psychiatry, 1957, 20, 11-21.

Starr, A., & Phillips, L. Verbal and motor memory in the amnesic syndrome. Neuropsychologia, 1970, 8, 75-88.

Symmes, D. Flicker discrimination by brain-damaged monkeys. Journal of Comparative and Physiological Psychology, 1965, 60, 470-473.

Tulving, E., & Pearlstone, Z. Availability versus accessibility of information in memory for words. Journal of Verbal Learning and Verbal Behavior, 1966, 5, 381-391.

Warrington, E. K., & James, M. An experimental investigation of facial recognition in patients with unilateral cerebral lesions. Cortex, 1967, 3, 317-326.

Warrington, E. K., & Rabin, P. A preliminary investigation of the relationship between visual perception and visual memory. Cortex, 1970, 6, 87-96.

Warrington, E. K., & Shallice, T. The selective impairment of auditory verbal short-term memory. Brain, 1969, 92, 885-896.

Warrington, E. K., & Weiskrantz, L. A new method of testing long-term retention with special reference to amnesic patients. Nature, 1968, 217, 972-974. (a)

Warrington, E. K., & Weiskrantz, L. A study of learning and retention in amnesic patients. Neuropsychologia, 1968, 6, 283-291. (b)

Warrington, E. K., & Weiskrantz, L. Amnesic syndrome: consolidation or retrieval? Nature, 1970, 228, 628-630.

Weiskrantz, L. Central nervous system and the organization of behavior. In D. P. Kimble (Ed.), The organization of recall. New York: New York Academy of Sciences, 1967.

Weiskrantz, L. Some traps and pontifications. In L. Weiskrantz (Ed.), Analysis of behavioral change. New York: Harper & Row, 1968. (a)

Weiskrantz, L. Experiments on the r.n.s. (real nervous system) and monkey memory. Proceedings of the Royal Society (London) Series B: Biological Sciences, 1968, 171, 335-352. (b)

Weiskrantz, L. A long-term view of short-term memory in psychology. In G. Horn & R. A. Hinde (Eds.), Short-term changes in neural activity and behaviour. Cambridge: Cambridge University Press, 1970, pp. 63-74.

Weiskrantz, L., & Cowey, A. Striate cortex lesions and visual acuity of the rhesus monkey. *Journal of Comparative and Physiological Psychology*, 1963, **56**, 225-231.

Weiskrantz, L., & Warrington, E. K. A study of forgetting in amnesic patients. *Neuropsychologia*, 1970, **8**, 281-288. (a)

Weiskrantz, L., & Warrington, E. K. Verbal learning and retention by amnesic patients using partial information. *Psychonomic Science*, 1970, **20**, 210-211. (b)

Williams, M. Investigation of amnesic defects by progressive prompting. *Journal of Neurology, Neurosurgery and Psychiatry*, 1953, **16**, 14-18.

CHAPTER 3

Some General Characteristics of a Method for Teaching Language to Organisms That Do Not Ordinarily Acquire It

David Premack

INTRODUCTION

The approach we have taken to the question, "Can an ape learn language?," and to the more important question which this presupposes, "What is language?," can be expressed in terms of two parallel lists. The first of these is a list of exemplars, the things an organism must do in order to give evidence of having language. The second is a corresponding list of training procedures, one for each exemplar. When a training procedure is properly applied it has the force of producing the associated exemplar (or it is not called a training procedure). The set of exemplars and corresponding training procedures is at least partly ordered; exemplars vary in their linguistic prerequisites, some being prerequisites for others.

This general approach—exemplars and recipes for teaching them—is applicable to more than language. It may be useful in all cases where the behavior is complex and where it is not possible to generate an exhaustive enumeration of the evidence from a formal theory. Which is to say, the approach is useful in those cases where the task is not only to produce an outcome but to explicate the outcome beforehand. For example, we might ask, "Can apes acquire conscience?," and be led swiftly to the question, "What is conscience?" In this case, too, a start is possible by listing actions that define conscience—like language, conscience is too complex to consist of only one action—and then attempting to devise a means of teaching each of them.

This approach is likely to utilize infrahuman subjects, although it need not. We do not require specifically animal subjects, but simply subjects who do not acquire, in some cases reacquire, the disposition in question in the normal course of events. Thus, a test of the putative training procedures for language is not best made with a normal child, for this subject will acquire language willy nilly (though we might show that the acquisition is more rapid or efficient when the training procedures are used). On the other hand, certain clinical human populations can be used as revealingly as animal populations. For example, languageless autistic children and global aphasic adults are populations to which we have begun to apply the language training procedures. In principle, this approach can be taken to any organism who has lost a rich capacity which it is premature to suppose is unrestorable in principle. For example, a physicist may be injured gravely, recover, but after recovery no longer prove to be a physicist. Is he restorable? Yes, if the behavior of a physicist can be reconstructed, and training procedures can be shown that will realize the reconstruction.

Difficult as the case of the physicist would be, consider the restoration of a poet. Which is to say, in all interesting cases the problem lies far less with the training procedures than with the explication. On the basis of what exemplars do we recognize a poet? In general, how do we avoid trivial exemplars such that even if the training were to succeed, we would still fail? Here at least we have the advantage that comes from tying training programs to the exemplars; inadequacies in the analysis should show up early.

In judging the outcome, we need to be on guard against deciding on the basis of form rather than function. The would-be restored physicist might solve problems and even offer innovative solutions, and yet not do so in a modal way. Would this disqualify the restoration? The problem is especially acute when the subject is animal, for there the likelihood of duplicating the human form is remote. But we will not call a thing "clarinet playing" simply because the music is produced by playing the instrument with the mouth (rather than some other orifice) and the hands (rather than some other appendage). Do blindfolded judges call it clarinet playing or even music? That is a better criterion. Moreover, in most of the interesting cases, the important "playing" is internal. In such cases we cannot decide equivalence on the basis of the form of the process, for too often we lack an adequate account of the process. For these several reasons, we begin in this frankly operational way: a list of (strictly) functional exemplars and procedures for teaching them.

In this paper, I will describe the procedures we have used for teaching some of the exemplars of language to Sarah, an African-born female chimpanzee, who was estimated to be between five and six years old when the study began. Since these procedures and results have been described elsewhere (Premack, 1970) I will not attempt a complete list but here will concentrate on failures, cases where, although the exemplar was finally taught, it was taught with difficulty, even perhaps with doubt as to whether it was taught at all. In all of these cases, it is possible to describe alternative approaches. Not yet tested, they nonetheless seem patently better than those that were used. The work described was, after all, essentially a pilot study, and it would be strange indeed if it were not possible to improve upon it. In addition, I will generalize on the characteristics of the training program, and make a few guesses as to what some of the critical factors may be.

PHYSICAL BASIS OF LANGUAGE

The language is analogous to reading and writing rather than to speaking and listening. Each word is a piece of plastic, varying in size, shape, color, and texture, that is metal-backed and capable of adhering to a magnetized board (see Fig. 1). There are no systematic relations between properties of the words and

FIG. 1. Each word is a piece of plastic varying in color, shape, size and texture. The pieces are metal-backed and adhere to a magnetized slate. Sentences are written on the vertical. Literal translation of the two sentences shown are: "Sarah jam bread take," and "No Sarah honey cracker take." (From Premack, 1970.)

their meanings. Thus, the word for apple is a small, blue triangle; the names of colors are not themselves colored; the names of shapes do not have the shape of their referent. Sentences are written on the vertical to avoid problems of laterality that might arise if sentences were written on the horizontal and the several parties to a conversation did not write on the same board. There are no explicit phonemes in the language, the most elementary unit being quite deliberately the word. We do not need phonemes (i.e., graphemes) because, first, working in the laboratory we do not need the thousands of words that are required for the mapping of a real world, and second, carrying through the basic functions of language does not itself hinge on a large vocabulary. The phoneme or grapheme may be viewed as a solution to the "large" vocabulary problem. Organisms neither generate nor store efficiently thousands of irreducibly different responses; instead, they generate 50 or so phonemes and then make thousands of different words by combining the phonemes. But we do not need thousands of words for the reasons stated. In dispensing with the phoneme, we cannot study questions of phonology, but we suffer no loss in either semantics or syntax which is where we wish to place the emphasis.

The plastic language offers three major advantages for training. First, since words are permanent rather than transient—the sentences are displaced in space, not time—it is possible to study language without a short-term memory problem. Without this provision, any failure would be ambiguous, subject to the possibility that the subject did not remember rather than did not understand the instruction. Second, the difficulty of any task can be modulated by controlling the number and kind of alternative words available to the subject at any moment in time. Since in this system the experimenter makes the words, while the subject merely uses them, the words available at any moment in time can be varied in number, kind, type/token ratio, etc., as the experimenter chooses. It is a drawback for training in natural language that the child can say any word that pops into his head. In the present system, we can restrict the subject's vocabulary to just one word, and in so doing, assure that he will answer any question asked him in the didactically most beneficial way. Later on, of course, we can enlarge the subject's available vocabulary as desired, and thus test the limits of his knowledge. Third, on the other side of the coin, the child, though capable in principle of saying whatever pops into his head, may be unable to produce the desired word; it cannot produce any words until it has passed through the complex motor learning that is required for the production of words. In contrast, sticking a metal-backed piece of plastic on a magnetized surface is a simple act, one the subject is proficient in almost from the beginning. As a consequence, training can take place in production as early as comprehension, which is a decided convenience since some words are trained more easily in one mode than the other.

QUESTION

The question will make a good starting point. In fact, it is introduced as early in training as possible because of the advantage it offers for the teaching of further language. We introduced the question in the context of *same-different*, which was itself one of the first concepts taught because it has no linguistic prerequisites. We first established that the animal was capable of making same-ness-difference judgments (that coincided with our own), and did so with a matching-to-sample procedure. That is, we placed before her two cups and a spoon and rewarded her for bringing the two cups together. Later we offered a large assortment of items, always in sets of three, and found that she was capable of matching not only the few items we had used to teach her the matching procedure but an essentially unlimited array of items.

If the acquisition of language is the mapping of existing knowledge (as is widely suggested, e.g., Vygotsky, 1962; Piaget, 1962) then teaching *same-different* should be straightforward. Two cups were placed before Sarah at a small distance from one another; she was given a piece of plastic intended to mean "same" and required to place it between them. On other trials, she was given a cup and a spoon, also set slightly apart, and required to place between them a second piece of plastic intended to mean "different." Did she form the appropriate associations?

The same material was presented as before—two cups on some trials, a cup and a spoon on others—but she was given both "same" and "different" and required to choose between them. She made four errors in 26 trials, none on the first five trials. Extensive transfer tests were then given to determine whether she could apply the words to items not used in training. She could indeed. In principle, Sarah could go about the cage, pick up pairs of objects, and label them same or different. Instances of sameness and difference which she was capable of recognizing before, she could now label as such. This, as opposed to any new concept, is what the language training should be credited with.

In these exercises, Sarah was already asked a question, though without benefit of an explicit interrogative marker. The same task undertaken with an English-speaking subject would almost certainly lead to instructions along these lines: "What is the relation between the two objects; are they the same or different?" But the only indication of a question which Sarah had so far was the implicit one of the space between the objects, into which she was to insert her answer, along with the fact that a trial did not end until she completed the construction by adding the third item.

Of the three standard linguistic devices for marking the interrogative, inflection, word order, or an interrogative particle, we chose the latter as being the simplest, both in the sense of involving the least change for the subject and of being most readily adapted to the present physical system. Henceforth, rather than writing

A   A,
"same" "different"
we wrote:
A   ?   A.
"same" "different"

The method of answering questions was simple: remove the interrogative marker and replace it with the appropriate alternative.

The variability in mechanical devices that languages use to identify a statement as a question tends to obscure the basically simple nature of what a question is. Any completable construction is a potential question. It becomes a question once it suffers one or more missing elements. That is the structural account. From the psychological point of view, we must add that a question

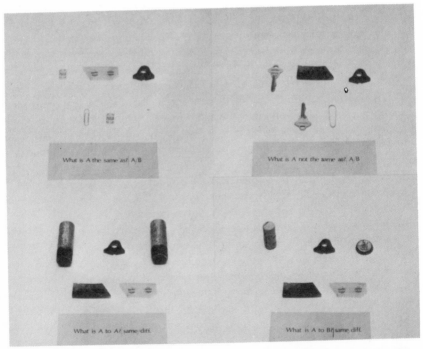

FIG. 2. Four *wh*-type questions with English paraphrases. Notice how the interrogative marker is substituted for objects instancing the predicates in the two questions in the upper panel, and for the predicates themselves in the lower panel. "Different" would be a more accurate paraphrase than the "not same" shown in the figure, since the negative particle was not used to form the negative case. (From Premack, 1970.)

arises when a speaker finds himself unable to complete certain constructions and has at hand a listener whom he regards as a probable source of missing elements. (Perhaps we should add that the construction need not lack exactly one element, but must not be so completely wanting in constituents as to be no longer recognizable as an incomplete version of a given construction.) If this treatment is correct, then the ontogenetically earliest context in which to introduce the question, and with great didactic benefit to all subsequent language training, is that offered by the simplest possible completable constructions. *Same-different* is such a construction since it can be introduced as a relation between unnamed objects and thus has no linguistic prerequisites.

Since questions rely on missing elements, with a two-term relation such as *same-different*, two question forms can be generated directly, one by removing the predicate ("same" or "different"), another by removing one or even both of the objects instancing the predicate (A or B). A third form can be generated indirectly by appending the interrogative marker, which itself stands for missing element(s), to the head of the construction and then requiring that it be replaced by a further element, specifically either "yes" or "no." Examples of all three questions are shown in Figs. 2 and 3.

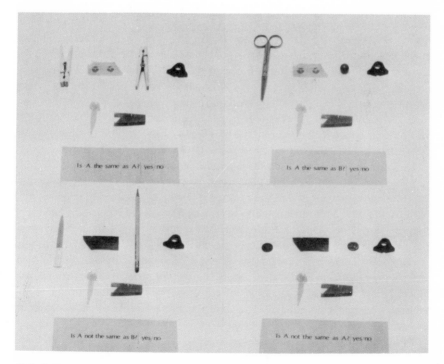

FIG. 3. Four *yes-no* type questions with English paraphrases. Again, a more accurate reading would be given by substituting "different" for the "not same" shown in the figure. (From Premack, 1970.)

An example of two versions of a *wh*-type question is shown in the upper panel of Fig. 2. These questions can be paraphrased as "X is what to X?" and "X is what to Y?" The alternatives are "same" or "different," and Sarah's task was to replace the interrogative marker with the appropriate word.

Two versions of a second type of *wh*-question are shown in the lower panel of Fig. 2; they can be paraphrased as "X is the same as what?" and "X is different from what?" Now the alternatives are no longer the words "same" or "different" but the objects themselves. Sarah's task remained the same, viz., to replace the interrogative particle with the proper object and thereby complete the construction.

The *yes-no* question, the third form that can be generated in this context, is shown in four versions in Fig. 3. They can be paraphrased as (i) "Is X the same as X?", (ii) "Is X different from X?", (iii) "Is X different from Y?", and (iv) "Is X the same as Y?" These questions were formed not by removing any item from the string, but rather by appending the interrogative marker to the head of the string.

The first two question forms involved only one new term, the interrogative marker itself. The use of this particle, its removal and replacement by the particle that completed the construction, was taught by direct intervention. Sarah was already thoroughly experienced in placing either "same" or "different" in the space between A and A, or A and B. The interrogative particle was inserted in the slot normally occupied by the predicate. Sarah's hand was placed on the particle and she was led to move it out of the way, leaving the customary A-A, or A-B, which she completed as usual with either "same" or "different." That constituted the formal training on the interrogative particle.

She was then given 20 questions of the second *wh*-type, in which one of the two objects rather than the predicate was replaced by the interrogative marker. On each trial, her alternatives consisted of two objects. She was slow to start work and was primed by again placing her hand on the particle and leading her to move it out of the way. She then chose correctly on the first six trials and on 11 of the remaining 14 trials. Aside from the "balk" on the first trial, which may have been a mild emotional reaction to a novel form, there was good evidence of transfer from Form one to Form two. The transfer cannot be ascribed specifically to the interrogative particle, however. She may have done as well without it. The interrogative marker in the *wh*-type questions is a redundant device, serving merely to give the blank space better definition than it would have otherwise. It is probable that the interrogative marker did contribute to the transfer. We simply do not have the control data that would make it possible to evaluate the weight of the contribution.

The *yes-no* type question involved four linguistic items—"yes," "no," the predicates "same" and "different," and the interrogative marker—and to teach this construction in the standard way it was necessary to assure that Sarah not be required to learn more than one new term at a time. Each training program seeks to teach its exemplar or concept in the fewest possible steps. In practice, the smallest possible step turns out (in most cases) to be equivalent to a word or particle (morpheme). A new word is rarely if ever introduced alone, but characteristically as an element in a string, a string in which all the other elements are known. The known elements may be nonlinguistic—actual objects—and thus known on a direct perceptual basis, or more characteristically, they may be linguistic elements that are known through prior training. Early in training the string in which the new term is introduced will necessarily be short, e.g., "A same A." Later on, strings may attain to considerable length; *if-then*, or the conditional particle, was introduced as the unknown between atomic sentences that were themselves highly familiar, e.g., "Sarah take red ⊃ Mary give Sarah chocolate" (Premack, 1971). But the length of the string is secondary. The important point is that in a strict training program there is only one unknown element. The introduction of one new element at a time minimizes the difficulty of learning the exemplar. In addition, if the subject should fail, the locus of the failure would be explicit as it would not be if two or more elements were introduced together. If a subject or species proved to be capable of learning many exemplars on a one-step basis, its capacity could then be tested by requiring it to learn other exemplars on a multistep basis. Once training programs have been devised which do not require more than one unknown to be learned at a time, it is always possible to test the limits, i.e., to complicate the problem and see how many unknowns the subject can learn at one time.

NEGATIVE

Of the several linguistic elements contained in the *yes-no* type of question, Sarah was already familiar with both the predicates "same" and "different" as well as the interrogative marker. These were taught in the exercises above. She was also familiar with "no," or the negative particle, by way of previous training that we have not yet described. "No" had been taught to her previously as a particle appended to the head of a construction with the force of an injunction against carrying out the action otherwise called for by the construction. Small pieces of bread and cracker were spread with peanut butter, jam, or honey and the cocktail party generated by this 2 X 3 factorial was arrayed before her in six columns of about five canapes each. On perhaps 20% of the lessons she seized the goods and the lesson proceeded no further. But on the other 80%, though the trainer looking at her across the table spread with all the goodies was powerless to prevent her, she did not seize the training materials but took only those she was instructed to take. (This obedience, which was not so much taught as generated by the social relation between the animal and her trainers, was an important part of the motivational basis of the training; that topic is discussed elsewhere.)

Pairs of sentences were written on the board before her. Initially both members of a pair were positive, e.g., "Sarah take honey bread," and "Sarah take jam cracker." On earlier lessons, the trainer had used the procedure of occasionally italicizing the new word in a construction by pointing to it, and Sarah may have been mimicking this gesture when she adopted the practice of pointing to the sentence she was attending to before reaching for the object referred to in the sentence. This made it easier to judge the correctness of her comprehension; if she had not indicated which sentence she was responding to, we could have credited her with a correct reading even though the object she chose was only coincidentally the same as one of the two referred to by the sentences. Of course, we could have required her to process the sentences according to a fixed order, say, left to right, but her use of pointing removed this necessity and at the same time made it possible for us to discover that she did not herself employ a fixed order. There was some tendency for her to choose a preferred item first, though her preferences among the six items were not strong, and the order in which she responded to the two sentences was not consistent. (In retrospect, her disinclination to use the same sentence twice was notable. After pointing to one of the sentences and correctly taking the item called for by the sentence, as she did about 80% of the time, she proceeded to the next sentence. She did not point to the same sentence again or return to the same item twice. Perhaps this was merely because a "new" item was always more interesting, though the pieces were deliberately very small. She was never given pairs of identical sentences, which could have been a hint not to process the same sentence twice, or otherwise taught to proceed from one sentence to the next. It is hard to say whether this behavior was the result of a "rule" of her own or the outcome of a simple preference for the "new" item.)

The negative particle was introduced by appending it to the head of one of the two constructions on some trials; on other trials we continued to give two positive or noncontramanded sentences. We debated inserting the negative particle into the sentence, e.g., "Sarah no take jam cracker," so as to direct its force specifically on the verb, but decided, I now think incorrectly, that initially the effect would be clearer if the particle were appended rather than inserted. (At this stage we were very hesitant about the certainty of her understanding and noted that to append would leave the base sentence intact while insertion would not. I think we attributed more fragility to her understanding than was called for. I dwell on this point because in appending rather than inserting we violated one of our own rules which was never to introduce quasi-acceptable forms of a construction which were later regarded as infantile and superseded by fully acceptable forms. That also is an objective of the training program. If particles are introduced as needed, there should never be a need to use quasi-acceptable forms that are later abandoned. In the present case, there was no such need and only our hesitation about her comprehension led us to violate the rule.)

"No" was taught her simply by staying her hand whenever she reached for an item referred to in a sentence to which a negative particle was appended. This simple procedure was extremely effective (as we have also found it to be with language-deficient children). The trainer frequently managed to catch Sarah's hand in flight. An arrested gesture is a highly discriminable event, thus, everything else equal, an event with which a word can be readily associated. After only about five or six such "arrests" she ceased even to point to the sentence to which the negative particle was appended and, of course, on those occasions did not reach for the item in question. A positive sentence was always given along with a negative one, as well as trials on which both sentences were positive, simply as a means of making obedience easier. (At later stages in the project, she disclosed several peculiar tendencies with regard to "no" and the subsequently introduced "none" which may or may not have related to the original manner of introducing "no." For example, at a later stage, she was required to answer yes-no-type questions concerning which of two cards the trainer placed on the other. Suppose the trainer placed a blue card on a red one, and asked her, in effect, whether a green card was on the red one. About 40% of the time, rather than answering "no," she replaced the blue card with a green one and answered "yes." That is, about 40% of the time, she modified the world so as to make a "yes" answer possible or a "no" answer avoidable.)

Rather than give her the usual transfer test, which consisted of new items being substituted for the training ones, we tested her on the other two forms of the yes-no question. The four forms of the yes-no question were intermingled and she was asked them in a more or less random order. Her alternatives were "yes" and "no" as before and the objects substituted for X and Y were those used in training.

On Questions 1 through 4 (see Fig. 3), she made the following number of errors per total number of trials: 6/33, 11/43, 2/27, and 11/51. Errors were concentrated in Forms two and four, both forms in which the word "different" appeared. She made 25 errors in 94 trials on "different" questions, only eight errors in 60 trials on "same" questions. Her error distributions were otherwise about equal. She made approximately the same number of errors on the two new forms introduced on the transfer test as on the two old forms (13/78 vs 20/76), and about the same number on questions requiring "yes" and "no" answers, respectively (17/78 vs 16/70).

She evidently did not learn the "same" question in the same manner as she learned the "different" question. The data suggest she learned the latter simply as a correlation between "different" and "no," i.e., write "no" whenever "different" is present. This simple rule failed, however, on the transfer test when the second form of the "different" question was introduced, for the correct answer to it was "yes." That is, on the original training it was sufficient to answer the "different" question with "no," the "same" question with "yes"; but on the transfer test "different" questions required both "yes" and "no" answers, as did "same" questions. This explains why after making virtually no errors on the "different" question in the original training, she went on to make numerous errors on this same question when the second version of the "different" question was introduced. Interestingly, although the "same" question was equally subject to the same kind of learning—write "yes" whenever "same" appears—she apparently did not learn it in this manner. The introduction of the second form of the "same" question did not occasion a large number of errors; she learned questions of this form at her usual level of proficiency. The data do not necessarily reflect an inherent difficulty in the *yes-no* question. They may also be seen as a judgment on a training program that was unsound because it allowed an inadequate rule of the above kind to develop.

BETTER PROCEDURES FOR TEACHING YES-NO QUESTIONS

How can the procedure for teaching the *yes-no* question be improved? Although it is important for certain purposes to describe the actual training of Sarah, it is equally important, for other purposes, to ask, "Is a strict training procedure possible in this case?" Can the concept be taught in a way that does not require the subject to learn more than one unknown at a time?

Simply by training all four forms of the *yes-no* question at the same time, we could forestall the development of the inadequate rule that Sarah used in the case of the "different" question. To do so would sacrifice the test of her ability to transfer from the first to the second two forms, which was the original point of training her on only two forms. But by teaching all forms together, an inadequate rule such as *Write "no" whenever "different" appears* would be nipped in the bud; one form of the "different" question requires "no" but the other form requires "yes."

A more interesting possibility is to extend the meaning of the negative particle in a direction more appropriate to its use in the *yes-no* question. The original training with the negative particle, e.g., "No take apple," etc., might be said to have given it a meaning close to "don't," a sense far removed from the one required in the *yes-no* question. Indeed, the original and presently required senses of "no" might differ so greatly that the subject would learn a new particle as "no" (in the *yes-no* question) as readily as she learned to adapt the old negative particle to this purpose. Only a test could settle this matter.

Negation rather than the original injunction first associated with the negative particle is the sense needed in the *yes-no* question. This sense could be taught

Sarah quite explicitly even at this relatively early stage of training. By making the word "different" temporarily unavailable, she could be required to write "A no same B" rather than "A different B." By a similar procedure she could be taught the corresponding equivalence between "same" and "no different." This sense of negation rather than being merely closer to the one required in the *yes-no* question is in all likelihood identical to it. Consider the two sentences, "No, A same B" and "A no same B." They might be paraphrased in English as, It is not the case that A and B are the same, and, A is not the same as B. The former would be generated in Sarah's language as an answer to a *yes-no* question, the latter as an answer to a *wh*-question (on the special condition that the word "different" was not available, or perhaps even with it available, if the subject had a preference for the negated form). The two sentences have the same meaning, establishing that the sense of "no" in the negation of a concept is the same as its sense in the *yes-no* question.

In fact, Sarah was taught how to form "different" by negating "same," and how to form "same" by negating "different," though she was not taught either equivalence until *after* being taught the *yes-no* question. Hence, we cannot ask, "Did training on negation facilitate learning of the *yes-no* question?" But we can ask the opposite, "Did use of the negative particle in the *yes-no* question facilitate the acquisition of negation?" Unfortunately, we must settle for a tentative or suggestive answer because of an unforeseen problem.

In teaching Sarah negation four questions were used, each with two alternatives. (i) "A ? A," with "same" and "no" as alternatives. (ii) "A ? A," with "different" and "no" as alternatives. (iii) "A ? B," with "same" and "no" as alternatives. (iv) "A ? B," with "different" and "no" as alternatives. She was not trained on any of the forms but tested directly on question forms (i) and (ii) to determine whether her use of "no" in the *yes-no* question would make it possible for her to form the equivalent of "same" by properly combining "no" and "different."

The test was ill-conceived or premature. Question (ii) required her to substitute two particles for one interrogative marker, an act for which she had no preparation. She was hesitant about making such a substitution and in the necessity of having to shape her to this act, we lost most of the information we were after. She should have been taught to make such substitutions in another context; the necessity of training in this act would not then have confounded the present test. After verbally encouraging her to add the second particle for several trials, we instated a time rule. Three seconds after presenting her with a question, we informed her whether her answer was correct or not, and did so for all questions in the exercise. For example, if in answering "A ? A" she inserted either "no" or "different" but did not add the second word within the 3-second time limit, she was marked incorrect (though the trainer was lenient on the first three or four trials).

After testing her on the first two questions, we added the other two and then asked her all four questions in a more or less random order. On questions (i), (ii), (iii), and (iv) she made the following number of errors per total number of trials: 11/79, 9/77, 5/37, and 7/35. Her errors on the first five of each of the same questions were: 1/5, 1/5 (on three of the five she did not meet the time limit but got the correct answer), 0/5, and 0/5. Although the data do not establish that she learned negation through her experience with the *yes-no* question, they do show that she was ultimately capable of negation, and also, that training on the production of "same" through the negation of "different" (question ii) transferred to the opposite case, the production of "different" through the negation of "same" (question iii).

Do the two senses of no—injunction and negation—overlap or should we introduce separate particles for each? Whether they overlap in a helpful way could be determined by comparing Sarah's learning to negate "same" and "different" with a new particle on the one hand, and with an old negative particle on the other, one whose history included use in either or both injunction and the *yes-no* question. The same comparison could be made between her learning of a new "no" in *yes-no* questions and of an old "no" that had been used injunctively. Moreover, since there is no reason why injunction must precede negation, the question could be asked in the reverse, does the use of "no" in negation facilitate its subsequent acquisition in injunction? Intuitively, we considered the injunctive and negational senses of "no" to overlap and therefore did not introduce separate particles, but we have no data on the matter. The larger issue to which this point relates is that of polysemy, the multiple meanings which most words in natural language have. The two senses of the negative particle was our most explicit case of polysemy.

STRICT TRAINING PROCEDURE: ONE UNKNOWN AT A TIME

So far we have equated the concept of unknown with a word or particle. For example, we have said that a string realized the objective of a strict training procedure if only one new particle was introduced by the string. The strings used to teach the predicates "same" and "different" realized this objective in a unique way; the two items that, along with the predicate, made up the string were actual objects, not particles in the language, and were considered to be known as such. A more typical string was the one used to introduce the concept "color of"; "Red color of apple," is an example of this kind of string. In this case there are no objects in the string, only linguistic elements. Both the words "red" and "apple" were known, making "color" the only unknown. The string thus qualified as a strict training procedure. In these examples and in most others, the unknown is readily identified; it consists of a word, the one word not known to the subject.

The *yes-no* question shows, however, that an unknown need not consist of a new word (or particle) but may also consist of an old word that is operated upon in a new way. A specific example is the use of the interrogative marker in the *yes-no* question. In both *wh*-questions, where the interrogative marker was introduced, the words used to replace the marker have the direct effect of completing the construction. In the *yes-no* question, words replacing the interrogative marker do not complete the construction, since it is already complete, but add to it. The physical act of replacing the interrogative marker with another word is the same in both kinds of questions, but the logical act is substitution with completion in the *wh*-questions and substitution with addition in the *yes-no* questions.

It seems proper to treat this kind of difference as no less an unknown that the subject must learn than the more common case of a word that is unknown and must be learned. We are less inclined to think of a change in the manner of operating upon a word as an unknown, because this language was largely (deliberately) restricted to a single operation, viz., the addition of a particle(s) to a string. Since with few exceptions words were all dealt with in this one way, we tend not to consider that language involves operating upon words and that even in the present case, there was occasionally more than one kind of operation.

For example, the word can be added to a string of one or more other words or set forth as the first member of the string. This is the main operation in the production of sentences and is of interest when the subject is required to choose among alternatives. Suppose a piece of banana is placed before Sarah, along with a set of words, and that she wants the banana. She must choose the words that will permit her to form one or more of the possible banana-getting sentences. For example, all of the following sentences might be made possible by the words available to her: "Sarah want banana," "Mary give Sarah banana," "Sarah like banana," "Sarah want fruit," or even, "Sarah like Mary." (I add the latter because one can picture the situation in which the words available to her do not permit either a direct or indirect request for banana—e.g., "Give Sarah banana," [direct request] "Sarah like banana" [indirect request] —but in which, of the sentences that are possible, "Sarah like Mary" is the one most likely to get her the banana. This use of flattery is an interesting case; what must a subject know in order to use language in that fashion?) In the simplest case, the available words will permit only one banana-getting sentence (though it is difficult to prevent fragments of that sentence which can also be banana-getting. One could actually associate different probabilities of giving banana with sentence fragments of varying length; her likelihood of using a full sentence or one or another fragment could then be predicted from the probability-delay of reward function, delay being computed in terms of the time it takes her to produce the sentence.

If her likelihood of error increased with sentence length, this could be taken into consideration by relating its effect to the probability of reward.)

What goes into the production of the appropriate sentence? The sentence can be produced in several ways.

Selection and Addition

She looks out over a "large" set of words and selects from the 15 to 20 present those four that will make up the sentence. Then from these four she builds the sentence. On a few occasions she has proceeded in this way. Once, she was asked a series of questions all of the *yes-no* form, so that the only words she was using were "yes" and "no." On this occasion she took "yes" and "no" from the pile and pushed the other words to the side. On another occasion, she was required to describe a simple physical arrangement that the trainer changed from trial to trial. The trainer had four differently colored cards and on each trial put one on top of the other. Sarah's task was to describe the situation by writing "red is on green," "green is on red," etc., depending on what was before her. On this occasion, too, she withdrew the "relevant" words from the 15 or 20 words before her—names of the four colors, the preposition, and the copula—pushed the others aside and did not return to them for the remainder of the lesson. It is not hard to guess the circumstances leading her to behave in this manner. First, the consistent use of a type of sentence or nonlinguistic state of affairs (e.g., X on Y) make it possible to anticipate the words that were likely to be needed. Second, the number of irrelevant and thus possibly interfering words was large, making it worthwhile to eliminate them. In most lessons neither of these factors applied. Frequently, she had less than 10 words available to her throughout the lesson and on at least some lessons was asked several kinds of questions.

Rearrangement of Order

In this written language, the order of production need not be the same as the order of the product. Notice that the grammar talks only about the order of the product. For example, "Mary give Sarah banana" is correct whether it is written in its present order or any other. That is, "banana" could be the first word written, then "give," then "Mary," etc. In spoken or gestural languages—in all languages in which the symbols are displaced in time—the production and product orders are necessarily the same. But in a language in which the symbols are displaced in space, these orders can differ. In the present case, the grammar could be stated in terms of either production or product order, or both (though to state it only in terms of the order of production would throw away the advantages to short-term memory of the written case). In fact, in the beginning her production and product order did not agree. Even though the trainer sitting beside her always produced sentences in exactly their final order, Sarah often put the correct words on the board in one order, and then made one and, on occasion, two changes in their order before settling on a final order. Moreover, both the original order and subsequent changes were generally consistent. Thus, "Mary give Sarah banana" was written "Sarah" first, "give" next and below "Sarah," followed by "banana" below "give," and then "Mary" put at the head of the sentence. She then interchanged the positions of "give" and "Sarah" resulting in the grammatically acceptable order. In this case, we see her using two different operations: adding words and changing their order. We can distinguish a third operation if we want to differentiate between adding to the "head" and "tail" of the sentence (the difference between her addition of "Mary" to the top of the string, and both "give" and "banana" to the bottom. Neither bottom nor top were actually head and tail, but merely temporary outlying points in sentence development. Even so, her usage might be seen as precursor of suffix and prefix.) At a later stage of development, Sarah abandoned this uneconomical manner of sentence production and for the most part produced sentences in their final order like the trainers. Occasionally she reverted to the early production method, presumably when stressed by a change in trainer or the invasion of persons with cameras (Premack, 1971).

Substitution

The simplest operation is substitution, specifically, one-to-one substitution. This was the operation used in teaching her most new words. We gave her a string of particles with one interrogative marker either in a terminal position ("? color of chocolate," in effect, what is the color of chocolate?) or a bounded position ("brown ? chocolate," in effect, what is brown to chocolate?) In the case of training, she was given a single word with which she replaced the interrogative marker. The operation was simple in three senses: there was no choice of words, no choice of location for the word, and the product was a complete construction. At later stages she succeeded in more complicated kinds of substitution that were not one-to-one. For example, she answered the question, "A ? B," by writing "A no same B," as well as the question, "red, yellow ? fruit," by writing "red, yellow no is pl fruit," (in effect, red and yellow are not fruit) in the first case substituting two words and in the second case three words for a single interrogative marker. In substituting several particles for one, she inevitably must decide not only what words to use, but in what locations.

Deletion

Deletion, which is the last operation, was used in the training on the compound sentence (Premack, 1970). She was first taught the individual sentences from which the compound was derived—"Sarah insert apple dish" and "Sarah insert banana pail." The individual sentences were then combined and the redundant elements deleted one at a time:

Sarah	Sarah	Sarah
insert	insert	insert
banana	banana	banana
dish	dish	dish
Sarah	insert	apple
insert	apple	pail
apple	pail	
pail		

She performed at nearly the 80% level when confronted with the fully transformed compound and new words were substituted throughout the sentence without impairment. Her experience with deletion was thus in the comprehension mode. She was never taught to delete elements herself. Nevertheless, we gave her a test in which the correct answer depended on deletion. In the course of a series of questions concerning use of the plural particle ("pl"), she was given the question, "red ? pl fruit," in effect, what is the (class) relation between red and fruit? The correct answer required two operations (without regard to order). Substitute "no" and "is" for the interrogative marker, delete the plural marker, "pl." She substituted properly but did not delete. Thus, she wrote, in effect, "red are not a fruit," rather than the correct "red is not a fruit." It is difficult to interpret a failure of this kind. We can regard it as proof of how dim her grammatical knowledge really is. I was tempted to draw a similar conclusion when she was required for the first time to substitute two particles for one interrogative marker; her hesitation was intense. But once she was shown this operation—shown that it was legitimate?—she applied it correctly to new cases (see section on negation, page 59). Thus, her failure to delete may mean merely that she does not remove words from sentences because, so to speak, she knows better. (It was some time before I could bring myself to underline words in books or write in their margins. High school rules had to be discarded before those of the University could be adopted.) The point is simple. We cannot make an unconfounded test of her grammatical knowledge with tests that require her to perform operations—many-one substitution, deletion—which she has never performed before. She must be experienced in the operation in some other setting before we require the use of the operation in a test of grammatical knowledge. The reader can easily guess why she was not trained, especially in the early stages, in either deletion or rearrangement. Either tendency would have disrupted the major training procedure, since this consisted of requiring her to insert one word into a string at a marked location. Allowing her to alter the string in any other way would have destroyed the viability of the simple training procedure.

A last method for teaching the *yes-no* question would be to regard substitution with completion and substitution with addition—the difference between the *wh-* and *yes-no* question forms—as no less demanding than the addition of a new word or particle. In this case, every word needed for the *yes-no* question should be introduced first, and the new manner of operating on the interrogative marker treated as the unknown. There is no problem in introducing any of the words needed in the construction with the possible exception of "yes." So far, at least, I have not seen a means of introducing "yes" except through the *yes-no* question.

In summary, we were largely successful in restricting training to one operation: one-to-one substitution, the introduction of one unknown element at a marked location in a string of known elements. Training took this limited form, but the more general use of language did not. Selection from a set of words, addition at unmarked locations, many-to-one substitution, rearrangement and deletion (in the comprehension mode) occurred successfully in some degree in her use of the language. The objective of a strict training program is the introduction of one unknown at a time. An unknown was almost always a new particle (word or morpheme) operated upon by one-to-one substitution. But on a few occasions, it consisted of an old particle operated upon in a new way.

PROPERTY NAMES

Consider the class concepts "color," "shape," and "size." We introduced each of them as the relation between specific properties and objects that instanced the properties, e.g., "red color of apple," where "color" was the new word. It was possible to use either objects or names of objects to instance the properties. We had an abundant supply of both actual fruits and fruits that were named. But we lacked the property names themselves, the color, shape and size names themselves, e.g., "red," "square," "large." How should we introduce the names of individual colors, shapes, and sizes, and of property names in general?

There are at least three methods that can be used, two of which we tried. One of the two failed for reasons we do not understand. I will describe all three methods mainly for purposes of speculating about the one that failed. Unfortunately, we cannot do more than speculate; in working with only one subject, it was difficult to eliminate the possibility that the failure of the one method was a necessary condition for the success of the other.

In the first method we offered Sarah pieces of apple dyed either red or yellow. The dye was tasteless and except for the difference in color the pieces of apple were identical. The training consisted simply of arranging that Sarah write "give yellow" on those trials on which the yellow piece was offered and "give red" on those trials on which the red piece was offered. After about 10 trials of each kind, she was given both the words "red" and "yellow" on each trial and required to choose between them. She performed at chance level and continued to do so after several hundred more trials of the same kind. Among the several possible causes of this surprising failure we gave priority to the fact that apple was an already named entity. She was being asked to write "give red," or "give yellow," for something which she had earlier learned to request by writing "give apple." We considered two alternatives: require her to write "give yellow apple" (or "give red apple,") introducing the to-be-learned property name as a modifier of an already established object name (which is the one method we did not try) or apply the same method but to objects that were not yet named.

We offered her cookies, an unnamed object at that time, which were either round or square, but otherwise the same. Again, the training consisted simply of requiring that she write either "give round" or "give square" on the appropriate occasion. She was tested in the same manner, i.e., given both the words "round" and "square" and required to choose between them. She performed at chance level and continued to do so after over 100 more trials of the same kind. The guess about interference through an existing name was clearly wrong.

Next, we considered that in locating the properties to be named in such objects, as apple and cookie, we may have led her to eat the properties without sufficiently attending to them. Since both pieces of apple tasted the same and both cookies likewise, she would have no preference between them, and thus little reason for closely attending to them. To make her more likely to attend to the difference between the apples, we attempted to use them as names rather than as referents for names. We offered her the piece of red apple as the name for gumdrop and the piece of yellow apple as the name for cupcake, two sweets that were not yet named. (In a system in which words and things are equally material, this kind of interchange is readily possible. Yet this was the only occasion on which we tried to use something other than the characteristic piece of plastic as a name. Nevertheless, we often found it convenient to use what we called "hybrid strings," i.e., strings consisting of both words and objects. However, in these cases, the object was never a name for anything else. For example, she was asked such questions as, "? color of" feather, in effect, what is the color of feather, where the feather was an actual object and one that was so far unnamed. The feather was placed next to the language particles "?" and "color of," in the same slot the word "feather" would have taken had there been such a word. So in this direct sense, the string, consisting of two words and a nonlinguistic object, could be viewed as a sentence. But the feather, unlike the linguistic particles, did not represent anything other than itself; i.e., if it were to be regarded as a name, it could only be as a name for itself. There is another possibility, however. When an object is placed in a string of this kind, it may represent not only itself, but the class of which it is a member. For example, suppose the subject were asked, "?" chocolate (the object) "is brown," in effect, is it the case that chocolate is brown? Even though the piece of chocolate used in the string was itself brown, the subject might nevertheless answer, "sometimes, and sometimes it is white," assuming now that her experience with chocolate included both colors, in which case, it would be clear that the piece of chocolate was representing not only itself but the class of which it was a member. Incidentally, if subjects are shown to do this sort of thing, to respond to an object as representing the class of which it is a member, it would be further evidence that language training itself does not teach organisms to use one item to stand for another, but that this is a primitive disposition which language takes advantage of, indeed, could not exist without.)

The training took essentially the same form as before. A gumdrop was placed before her on some trials along with the words "give" and "gumdrop" (the latter consisting of the piece of red apple), and she was to write "give gumdrop." On other trials with the cupcake present rather than the gumdrop, she was to write "give cupcake," using the other piece of apple as the name for cupcake. After 15 trials of each kind, she was tested by being given both would-be words and required to use the one appropriate to the object that was present. She performed at chance and continued to do so after more than 50 additional trials.

It was now necessary to determine whether she was color blind (though the failure on the shapes suggested that this was not the problem), and we did this by again using the red and yellow pieces of apple as cues rather than referents. This time, however, we used them in the simple role of discriminanda in a simultaneous discrimination problem. The two pieces of apple were set before her; when she chose the yellow one, she received nothing, but when she chose the red one, she was given the words "Sarah, banana, give," with which she could write the sentence "give Sarah banana," and for which she was then given a piece of banana. (We did not want the piece of red apple to become a synonym for "banana," so we gave her words for this choice rather than the banana itself. Was this reasonable? How do we know that the red apple did not come to mean banana anyway, or perhaps words, or even words that are banana-getting? In fact, we don't. But they seemed less likely alternatives than that red apple would become a synonym for "banana" if the fruit were given in direct exchange for the piece of apple.) She learned this problem in four or five trials, proving that she was not color blind and that she could discriminate between the two pieces of apple.

Using the pieces of apple as would-be names is comparable to a conditional discrimination problem—red if gumdrop, yellow if cupcake—and is more complex than their use in the simultaneous discrimination problem. There, one was "correct" and the other "incorrect." But all the words she was learning in this same period could be equally well viewed as conditional problems. They differed from this case only in that the referents differed by more than a single property.

Following the remarkable failure based on the use of objects identical except for the properties to be named, we adopted an essentially opposite approach. It had seemed a good idea to offer objects differing only in the properties to be named. What better way to assure that nothing but the desired properties would become associated with the words than to eliminate all other differences? Until it failed, it seemed a perfect procedure. The alternative we tried was to use a set of objects having nothing in common except the property to be named. For example, in teaching "red" and "yellow," we used a set of red and a set of yellow objects; every object in both sets was completely unlike every other one except for the common property of color. Thus, the red set consisted of a ball, toy car, lifesaver, etc., and the yellow set of a block, crayon, flower, etc. (None of the items in either set was named, though we have no evidence that that was a necessary condition.) Notice that in the first approach we eliminated all differences but one, forcing her, we thought, to observe that the pieces of apple differed in color. In the second approach, by allowing the objects in each set to vary in a number of ways, she was allowed to discover that their common property was color. It is not self-evident that the latter would be the successful approach.

The objective was to teach her to request each red item with the sentence "Mary give Sarah red," and each yellow item with the sentence "Mary give Sarah yellow." After requiring her to write each sentence at least once for each of the six items in each set, she was tested by being given both the words "red" and "yellow" and required to choose between them. That is, if a red item were present, she was required to write ". . . give red," and if a yellow item, ". . . give yellow." She passed the test at her customary 80% level and did equally well on a transfer test in which the procedure was exactly the same except for the use of red and yellow items not used in training. Finally, she passed the same test with respect to identical 2 X 4 cards painted red and yellow, respectively, thereby succeeding in what she had been unable to do earlier. That is, apply different words to objects that differed only in their color. The method was equally effective in teaching her the words "round" and "square," and later "large" and "small." Two sets of objects were assembled that had nothing in common except that in the first case all members of one set were round and all members of the other set square. (And in the second case, all members of one set were large and all members of the other set small.) The training consisted merely of requiring her to write ". . . give round" and ". . . give square" as well as ". . . give large" and ". . . give small" on appropriate occasions. She learned the proper associations and passed the transfer tests in these cases as in the case of color.

The method of sets should better prepare her for the transfer test than the method of single differences. The latter offers the red-yellow difference in only one context whereas the former provides the same difference in a variety of contexts. But there was no opportunity to test this reasonable hypothesis; she did not learn to name the red-yellow difference at all in the one case and passed the transfer test perfectly (i.e., performed at the same level as on the training items) in the other case.

How can we explain the success and failure of these two methods? Notice that the method of sets allows for the explicit disconfirmation of incorrect hypotheses, while the other method does not. If she should call the large car "red" and the small crayon "yellow," supposing the difference to be one of size, she would probably be disconfirmed on the next trial. "Yellow" is not the correct word to use with the small but red lifesaver, no more than "red" is correct with the large but yellow block. Differences in shape and indeed all differences save the one of color will undergo the same fate. In brief, if she entertains hypotheses and proceeds in this way, then with the method of sets, she can try the hypotheses and disconfirm them explicitly.

With the method of single differences she cannot explicitly try and disconfirm incorrect hypotheses. But the whole point of the method was to prevent the emergence of an incorrect hypothesis, and thus avoid the necessity of disconfirming it. So, although we must admit that the method does not provide for the explicit disconfirmation of incorrect hypotheses, it is not clear why she should ever entertain an incorrect hypothesis to begin with. For example, the pieces of apple did not differ in size; is there any reason, therefore, why she should have entertained a size-difference hypothesis in the presence of the apples? Unfortunately, if we answer "no," and generalize the answer, we cannot explain the failure of this method. For, while it contains no basis for disconfirming incorrect hypotheses, it would also seem to provide no basis for incorrect hypotheses to occur in the first place. The very failure of the method forces us, then, to consider the possibility that hypotheses are not always generated by physical differences. For example, despite the fact that the apples did not differ in size, should we consider that the subject might entertain this hypothesis anyway? But why and how? and which apple would she consider to be the larger one? The account is unclear at every critical point. All that one could say in defense of this argument is that if the subject should entertain an incorrect hypothesis, the method of single differences is not self-corrective; it provides no way of eliminating false hypotheses.

CLASS MEMBERSHIP

In teaching Sarah a word to denote class membership, we did not make adequate allowance for the abstract character of the relation. What is meant in saying that class membership is an abstract relation? Compare, for example, color as the relation between red and apple with class membership as the relation between apple and fruit (or red and color). In the case of color, it is possible to provide referents for both terms in the relation, but in the case of class membership one of the terms is elusive. In saying, X member of Y, X is easy to provide a referent for, but Y is not. That is, the class member is easily instanced, but the class itself is not. (Should we treat a list of class members as a referent for the class, and thus introduce class membership as the answer to the question, What is the relation between, say, "apple" and "apple, banana, orange, and raisin"?) In teaching abstract relations, any referent that cannot be provided in the usual direct way needs to be replaced by a functional substitute. I will describe the inefficient procedure we actually used as well as a better one for future use.

We gave Sarah the question, "red ? color," in effect, what is the relation between red and color? The only alternative given her was "is," with which she replaced the interrogative marker, forming the sentence "red is color." Next she was given the sentence "round ? shape," in effect, what is the relation between round and shape? Again, the only alternative given her was the word "is," with which she replaced the interrogative marker, forming the sentence, "round is shape." Following the usual five trials on each of the two positive instances she was given the same number of trials on each of the two negative instances.

Thus, she was asked "red ? shape," in effect, what is the relation between red and shape? as well as "round ? color," in effect, what is the relation between round and color? In both instances, the alternative given her was "is-not," i.e., "is" glued to the negative particle. She displaced the interrogative marker forming on one occasion the sentence, "round is-not color," and on the other occasion, "red is-not shape."

At Step two, she was asked the same questions but given both the words "is" and "is-not" and required to choose between them. She made 10 errors on the first 22 trials, all the result of a failure to use the negative form. In answering all questions with "is," she formed such blatantly erroneous sentences as "round is color" and "red is shape." Following this rare event, an unqualified failure at Step two, she was returned to step one, given ten more trials, five on each of the "is-not" or negative cases, and advanced again to Step two. This time she made 14 errors, of them 11 on "is-not." She was returned to step one, given only two trials on each of the negative cases, and retested on Step two-type trials. On this third and final test, she made no errors in 18 trials.

On the transfer test, she was asked all the previous questions with the words "yellow" and "triangular" substituted for the training words ("red" and "round"), and required to choose between "is" and "is-not" as before. For example, she was asked, "yellow ? shape," in effect, what is the relation between yellow and shape? She answered correctly, replacing the interrogative marker with "is-not," forming the sentence, "yellow is-not shape." She made only three errors on 26 trials, none on the first five trials. Thus, even though "is" was acquired with many errors, it was successfully transferred to nontraining items.

Why use property classes rather than object classes to introduce class membership? There were two reasons. Both reasons are unimpeachable so long as the problem is looked at in a certain light. But once the light is changed, the reasons are seen to be subvertible.

When the problem is looked at from an excessively rational framework, it appears that class membership could be introduced in a way that did not require more than one unknown, only with the use of property classes. In the training sentences that were actually used, e.g., "red is color," both "red" (the class member) and "color" (the class) were known. Only "is," the term denoting class membership, was unknown. A comparable state of affairs did not exist in the case of object classes. Not only did we not have object classes (unlike the property classes which we did have), but the impression was that we would need the concept of class membership itself in order to introduce object classes.

"Color" and the other property classes, "shape" and "size," were introduced as the relation between a property and an object instancing that property. For example, "red color of apple," "round shape of cracker," etc. But there was no parallel procedure that could be used in the case of object classes. While "color" can be introduced with "red color apple," there is no parallel construction that will introduce object classes. If we try to use "fruit" as a parallel to "color," thus saying, "____ fruit of apple," there is nothing to put in the blank. We might put "fruit" in the position of "red," i.e., treat "fruit" as a property and say, albeit awkwardly, "fruit class of apple." But to introduce either "fruit" or "class of" with that dubious construction would still require finding an independent way of introducing at least one of them, since they are both unknowns. There are some other alternatives, but all equally unacceptable, and I will not consider them here. In brief, from a rational point of view, two conditions appeared to hold: (i) the relation between properties and a property class was the logically necessary way of introducing the copula or whatever term was used to denote class membership, and (ii) object classes could not even be introduced without the use of the concept of class membership (and thus could hardly be used to teach it).

The second reason for assigning a priority to property classes is because they successfully avoid the indeterminacy that holds between object classes and their members (this is correct but the implication I thought it had for training is not, which is what I mean by sound reasons that are subverted by a change in point of view). For example, with the use of "is" Sarah was taught three object classes, "fruit," "candy," and "breadstuff," a class invented for the occasion. After teaching her these classes with a small set of exemplars, e.g., "apple is (a member of the class) fruit," "chocolate is candy," "cracker is breadstuff," etc., she was given transfer tests requiring that she sort, among the three classes, cases that were not used in training. Thus, she was required to say, in effect, whether peach, a nontraining item, was fruit or candy, and whether cookie, another nontraining item, was fruit or breadstuff. She gave her customary performance on the transfer test.

But what meaning does failure have in this case? Having been taught that banana is a fruit, suppose she chose not to call peach or grape a fruit? In what sense could she be considered to be in error? Actually, in the present exercises she was not asked *yes-no* questions which would indeed have required her to decide whether or not peach, for example, was a fruit. Instead, she was merely asked *wh*-questions, requiring her to decide only which a peach was more like, a fruit or a breadstuff. Nevertheless, errors are difficult to interpret in the case of object classes. An item cannot partly be and not be a color, or partly be and not be a shape, or partly be and not be a size. Yet every object class is subject to this ambiguity. A stone that barks or a dog that never moves—are they animate or inanimate? Intermediate items can always be proposed in the case of object classes (e.g., banana bread in the case of fruit and breadstuff.) Notice, however, that there are no items intermediate between colors and shapes, shapes and sizes, sizes and colors. You may waver in deciding whether to call black and white colors, but whatever you decide on that point, you will have no tendency to classify either one as a shape or a size.

Because object classes are the victim of intermediate cases, whereas property classes are not, we assigned a higher priority to the latter, used them to map class membership, and then tried to use the mapping to set up object classes. Though logically defensible perhaps, it was tutorially inefficient, and we have devised a simpler procedure for future use.

After teaching the names for a number of fruits, candies, and breadstuffs, the individual names should be temporarily removed, and the new words "fruit," "candy," and "breadstuff" made available in their place. Then there would follow a series of sentences in which Sarah would write, ". . . give Sarah fruit," and receive, over trials, all members of this class. Similarly, by writing, ". . . give candy," she would receive all members of this class, and by writing ". . . give breadstuff," all members of this class. Choice tests could be made in the usual way, and transfer tests as well. For example, offer her a fruit, candy or bread-stuff that was not used in training, give her all three class words, and observe whether she used the proper one.

The ostensive definition, which is easily realized with this system where words can be temporarily removed, forcing the subject to adopt alternatives, would seem definitely to contribute to the notion of class. The procedure should contribute to a functional equivalence between the class and its members. "Give me fruit" would have the same effect over the course of trials as "give me apple," ". . . banana," ". . . orange," etc. With class defined ostensively, it seems likely that class membership could be mapped in the same manner we used with property classes, but with greater efficiency, even perhaps with no more than the normal frequency of errors.

SYMBOLIZATION: WHEN IS A PIECE OF PLASTIC A WORD?

When does a piece of plastic cease to be a piece of plastic and become a word? We might answer by saying, "When it is used as a word: when it occurs along with other words of appropriate grammatical class in sentences, and when it occurs as the answer or part of the answer to questions." For example, we consider a small piece of blue plastic to be the name for apple because (i) it is the word used when, for example, the subject requests apple, and (ii) it is the answer given when the subject is asked, "What is the name of apple?" This is a standard answer, and we cannot improve upon it, though we may be able to add to it. We might say in addition that the piece of plastic is a word when the properties ascribed to it are not those of the plastic, but are those of the object designated by the piece of plastic. By what means can we determine whether this condition obtains?

This can be done most directly by using matching-to-sample once again, this time to obtain independent features analyses of both the word and its referent. A features analysis of the apple can be made by giving a series of trials on which the subject is presented with the apple and a pair of alternatives. On each trial she was required to indicate which of the alternatives she considered to be more like the apple. The alternatives we used in the analysis with the chimp were: red vs green, round vs square, square with stemlike protuberance vs (plain) square, and round (no protuberance) vs square with protuberance (see Table 1). The alternatives could be words, if the subject's vocabulary permits, or objects instancing the properties named by the words as in the present case. That is, the subject could be required to decide whether the apple is more like the words "red" vs "green" or more like a red patch vs a green patch. Our use of the latter approach was dictated by Sarah's presently limited vocabulary.

After obtaining a features analysis of the apple, we repeated the test exactly except for replacing the object apple with the name for apple. Once again the subject was required to indicate whether the sample—now a piece of blue plastic—was, for example, red or green, round or square, etc. Although the sample was no longer a shiny red apple but a piece of blue plastic, the subject assigned

Table I

Features Analyses of Apple and "Apple"

+	–	Object			Word		
(R)	(G)	+			+		
○	□	+			+		
⌂	□	+			+		
⌂	○	–			–		

to the plastic the same properties she earlier assigned to the apple (see Table 1). Surely if we did not know that the plastic stood for apple, we would be confused by her analysis of it; we might reasonably conclude that she did not understand matching-to-sample. But this is ruled out by her analysis of the object apple which accords nicely with the human analysis. The properties she assigned to the word are immediately sensible if we consider that her analysis of the word was not of its physical form but of that which the form represents.

If we regard this outcome as evidence of symbolization, on the grounds that a word is a symbol, it becomes important to determine the origins of the process. Was it instilled by the language training procedure, perhaps by the use of language to obtain certain outcomes, or by some other part of the overall procedure? If we adopt what may seem to be this reasonable view, then at some point we must be able to show how, with an organism that does not symbolize to begin with, it is possible to teach it to do so. This strikes me as an overwhelming challenge, one well worth trying to avoid. One way to avoid this challenge is to adopt the view that symbolization is a general characteristic of learning rather than a by-product of language. That is, rather then view symbolization as a consequence of language, it may be more sensible to view the acquisition of language as being made possible by the fact of symbolization.

Although we need not hold that symbolization is a universal property of learning, it is interesting to take this view for several reasons. First, the evidence required to confirm the view is less radical or counterintuitive than it may seem at first glance. Second, the view can be shown to be virtually equivalent to Tolman's (1935) theory of learning.

In the context of a standard learning experiment, positive evidence would consist of the following: the subject would ascribe to the discriminative stimulus largely the same features it ascribed to the reward or punishment associated with the discriminative stimulus. Suppose a pigeon were trained on a multiple schedule in which the food in one component could be described as small, red, hard, and round, and the food in the other component as large, yellow, soft, and square. A vertical line is the discriminative stimulus in the first component, a horizontal line in the second component. What features would the pigeon ascribe to the foods if, for example, a procedure like the one described above for Sarah were adapted to the bird so that the question could be answered experimentally? Notice that assumming symbolization to be an integral part of learning does not require an answer to that question. How the bird represents the foods to itself is a matter for test. All we are required to predict is that whatever features the bird ascribes to the foods it will also ascribe to the discriminative stimuli associated with the foods. It is not necessary or even likely that *all* the features in terms of which the bird discriminates the two foods will be ascribed to the associated stimuli. The features ascribed to the stimuli need only be sufficient to allow the bird to predict from the set of possible objects the one that is associated with the stimulus. Even though, unfortunately, the above experiment is hypothetical at this time, it is worth considering because of the perspective it gives to infrahuman intelligence and the problem of language.

GENERAL COMMENTS ON TRAINING METHODS

We cannot state the necessary and sufficient conditions for teaching the various exemplars of language. This was a pilot study. Therefore, we concentrated on introducing the exemplars in whatever way possible, and made changes in training only when forced to do so by failure. If we had been totally successful in each and every case, we might have succeeded in our objective of teaching language to an ape while at the same time learning nothing about the necessary and sufficient conditions. Choked by complete success, we could only have gasped that some part of what was done was sufficient and even perhaps necessary. Failure can be more instructive, especially failure that is followed by success once proper changes are made. We failed often enough to have an idea of what some of the critical factors may be. In this section, I will (i) summarize the features that applied to most of the training programs, (ii) make some guesses as to what some of the critical factors are, and (iii) note some properties that the subject, rather than the training program, contributed to the final outcome.

1. In many though not all cases, the concept was taught by offering two positive and two negative instances of it. This was always the case when the concept was a two-term relation, e.g., "name of" and "color of," but also when it was less easy (though not impossible) to view the word as a two-term relation, e.g., "none" and "all." In general, the negative instances were formed by using the opposite pairing of the same arguments that appeared in the positive instances. For example, the two positive instances in "color of" were the pairs "red"–"apple," and "yellow"–"banana" (e.g., "red color of apple"); the negative instances were the opposite pairings, "red"–"banana," and "yellow"–"apple" (correct pairings in certain worlds, but not in Sarah's). Similarly, in "name of," the positive instances were "apple" and apple, i.e., the word and the object, as well as "banana" and banana; the negative instances were again the opposite pairings, "apple"–banana and "banana"–apple.

1.1. In most cases, we introduced two new words at a time. Of the two words, one might be (i) the negation of the other, e.g., "name of" vs "not-name of," "color of," vs "not-color of," (ii) the opposite but not the negation of the other, "same" vs "different," "large" vs "small," or (iii) neither of the above but words that were semantically related, typically by class membership, e.g., "red" vs "yellow," "apple" vs "banana."

1.2. We did not introduce pairs of words that were semantically unrelated, or, to put it more accurately, whose relation was weaker than the one shown above (since presumably most words are related in some degree). Thus, we did not introduce such pairs as "apple" and "red," "none" and "blue," "big" and "round," etc. There are two disadvantages in the use of such pairs: (i) it is more difficult with such pairs to lead the subject to attend to the appropriate property, and (ii) the negative instances generated by combining such pairs produce statements in which predicates take illegitimate, and not merely incorrect, arguments. Consider examples of both.

1.2.1. In teaching the subject "red" and "yellow" we collected two sets of objects, the members of both sets being highly dissimilar except for the common property of red in one set, yellow in the other. The procedure consisted simply of requiring the subject to write ". . . give yellow" when a yellow object was present, ". . . give red" when a red object was present. It was successful in this case and in the case of other properties as well. Suppose, however, we had tried to teach "red" and "square" together rather than "red" and "yellow." How would the two sets have to be composed such that the subject would learn to apply one word to redness, the other to squareness? The red set would have to include objects of all possible shapes except square, and the square set objects of all possible colors except red. Otherwise the subject could divide the sets on a number of grounds other than the desired one. For example, without the safeguard above, the subject might divide the sets as colored vs not-colored, or alternatively as square vs not-square. Probably it would be inclined to do the former if all members of the square set were either all one noncolor (e.g., black) or a variety of achromatic shades; and to do the latter if all members of the red set were either all one nonsquare shape (e.g., circle) or a variety of nonsquare shapes. Needless to say, I do not know what properties the subject would actually abstract in these cases. But consider the problem from the experimenter's point of view. If he tries to contrast "red" and "square," the sets necessary to insure the abstraction of the desired properties are more demanding than those needed if he contrasts "red" and "yellow" (or "round" and "square"). Is it reasonable to anticipate an appreciable overlap between what the experimenter and subject must learn? If so, the subject's difficulty in mastering a lesson will be proportional to the experimenter's difficulty in arranging the lesson. This, of course, is not only a mere rule of thumb, but an untested one.

1.2.2. When the two words are not adequately related the negative cases which they generate can be unsatisfactory. For example, suppose that "color of" and "shape of" were contrasted rather than contrasting each of them with its negation (e.g., "color of" vs "not-color of"). Instead of starting out with, say, "red color of apple" and "yellow color of banana"—two positive instances of the concept "color of"—we might write, "red color of apple" and "round shape of apple." Opposite pairings in the first case would provide "red not-color banana" and "yellow not-color apple," both sensible statements. But opposite pairings in the second case would result in "red not-shape apple" and "round not-color apple." True statements, but misleading ones. "Round" is not a legitimate argument of "color" nor "red" of "shape." A speaker who knew the meaning of these words would not use them in this fashion. Both imply improper conclusions, viz., that red might be the shape of, and round the color of, something other than apple. If a subject did make such statements, we would immediately want to know whether he would entertain the improper implications above, and thus be shown not to know the meaning of the words, or whether he could be led to add, "because red is not a shape and round is not a color." But in training a subject, we should not cause it to write statements that would be made only by a philosopher or by someone who did not know the language. The contrast between inadequately related words tends to produce improper negative instances.

1.3. Are both positive and negative instances necessary? Obviously, we are not in a position to say. Nevertheless, when we attempted to omit part of the usual training, she failed, and when we replaced the missing part, success followed. We taught the subject the quantifiers "all" and "none" by using sets of crackers which were all of one shape or another (Premack, 1971). In the presence of a set of five square crackers, we gave the subject this question, "?

cracker is pl square," in effect, how many crackers are square ("pl" is the plural marker such that "is" + "pl" equals "are")? The answer given the subject was "all," the new word being taught and the only word in the string that was unknown to the subject. Replacing the interrogative marker with the one word given her, she wrote "all cracker is pl square," in effect, all crackers are square. Next we confronted her with a set of five round crackers. But rather than arranging for her to say of them, "all cracker is pl round," in effect, all crackers are round, which would have constituted the usual second positive instance, we skipped this case and rushed her into a negative instance. In the presence of the set of round crackers, we induced her to write, "none crackers is pl square," in effect, none of the crackers are square. So in the presence of the square crackers she was led to write, all crackers are square, and in the presence of the round crackers, none crackers are square. When tested in the usual way, by being given both words "all" and "none," and required to choose between them, she performed at chance. The reader will observe that we left out both the second positive instance, all crackers are round, and the second negative instance, none crackers are round. When this information was added, the subject went on to learn. Actually, for emphasis we added a third set of all triangular crackers and in association with it a third positive and negative instance, viz., all crackers are triangular and none crackers are triangular. The third case was superfluous and even perhaps harmful. There may be an optimal number of positive and negative instances to use in teaching new words.

2. A perceptually salient referent appears to be highly desirable here as in all learning. Referent is probably better understood as a psychological state than as the external state that contributed to the psychological one. Although a simple operationism is always preferable, in certain cases, such as the interrogative, it seems doubtful that the referent can be adequately defined in terms of obtaining stimuli. The best definition I can suggest for the referent in the case of the interrogative is "missing element(s)." This assumes that enough of a construction is present for the subject to recognize that a part of the construction is missing; also, that the subject knows the kind of thing that is missing so that it will accept certain elements as possible answers while rejecting others.

2.1. A referent may be nonsalient, and thus lead to slow learning for either of two essentially opposite reasons. In one case there are too many possible referents and the training fails because it is not the desired referent, but another one that comes into association with the word. This problem is best seen when the subject is asked to learn on an observational basis. Then like parents, the trainers may "converse" in the presence of the subject arranging to use certain words in conjunction with certain activities. For instance, on one occasion the prearranged conversation included the word "pour," in conjunction with an activity that included opening a milk carton and pouring from it into a baby bottle which the subject was subsequently given. The training need not have been executed as poorly as it was. Yet only after the first attempt did we recognize how numerous were the perceptual highlights of the activity that we hoped would come into correspondence with "pour." There was the opening of the plastic carton and the shaping of the spout (an almost certain perceptual highlight in view of the animal's marked interest); the grasping of the baby bottle in the right hand, the milk carton in the left hand and the bringing of them into position in the frontal plane; the actual pouring which included the visible rise of the milk in the bottle (also viewed closely by the attentive animal), the reading of the miniscus, etc. All of this was called "pour." Of course, the situation could have been improved dramatically. Nevertheless, to arrange that the conversation keep pace with the activity, so that the word "pour" occurred in perfect conjunction with the desired activity was no simple matter.

2.2. The second case is the opposite in that there is no referent rather than a superfluity of them. This was almost certainly the problem in our original attempt to map class membership. When she was required to write "red is color" and "round is shape," in an attempt to teach her the copula, there was no nonlinguistic state of affairs which this sentence mapped. The negative instance did not improve matters, e.g., "round is-not color" and "red is-not shape." Nor was there even a cue inside the sentence, as there is for the use of the plural, telling her when to use "is." Instead, "red is color" and "red is-not shape" had to be learned more or less in the manner of paired associates. Learning finally occurred, but only after considerable drill and with many more errors than usual.

2.2.1. Perhaps the proposed improved method of teaching class membership could be generalized and shown to be helpful in all so-called abstractions. When writing "apple is (a class member of) fruit," "is" being the unknown, thus the one you are attempting to teach, there was no nonlinguistic state of affairs to point to as class membership. We propose to surmount this problem by arranging that "fruit" serve as the functional equivalent of "apple," "banana," "orange," "apricot," etc., and "candy" as the functional equivalent of "caramel," "chocolate," "gumdrop," etc. By making the names of the particulars temporarily unavailable, and arranging that the new word "fruit" serve to obtain what all the individual words earlier obtained, and making comparable arrangements in the case of "candy," both class words may be shown to become functional equivalents for their respective members. They will be used in place of their members if the members are unavailable, even perhaps when the members are available, provided the subject does not have marked preferences within a class, preferences that it can exercise by using class member but not class words (on the other hand, it may be more likely to err and thus lose out when using individual rather than class names). In brief, the subject has written ". . . give fruit" on all those occasions when it would otherwise write ". . . give apple (banana, orange, etc.)," and written comparable things in the case of candy and its membership.

Now presumably, the attempt to teach class membership by writing, "apple, banana, orange, etc. is pl fruit" and "caramel, chocolate, etc., is pl candy," will be more successful. Now the negative instances—"chocolate, etc., is not fruit" and "apple, etc., is not candy"—may be more meaningful. "Fruit" has not served in place of "chocolate," etc., nor "candy" in place of "apple," etc. That is, there are words for which "fruit" has substituted and words for which it has not; the same is true for "candy." This is one meaning, albeit a decidedly linguistic one, which class membership could be given. Namely, a,b,c,d,. . .n is a member of X, means that X can be substituted for any of the lower case letters with the same outcome they would have if used in the otherwise identical sentence. (If a piece of, say, apple is visible, both "give apple" and "give fruit" will have the same outcome; if the fruit is not visible, "give apple" and "give fruit" may have different outcomes, i.e., "give fruit" may result in banana, though this is less different than the difference between either of them and the outcome of "give candy.") An obvious alternative to this linguistic view is the more perceptual one in which an appeal is made to the properties that all members of the fruit class share, all members of the candy class share, etc. But this perceptual view is far more impressive with regard to property classes—colors, shapes, sizes, etc.—than to object classes.

2.3. The referent need not be a condition of the world but can be a condition of the subject as in the case of psychological terms. The only two psychological terms we have taught her so far are "prefer" and "want." They are easier to teach than, for example, "think" or "know," although it is not clear that the latter introduce qualitatively new problems. "Prefer" was especially easy to teach because the condition which it maps can be both measured and manipulated easily. We started by using choice tests to establish Sarah's preferences. The tests showed that Sarah preferred candy to all other foods and also had preferences among the fruits. She was taught the word "prefer" with the string "Sarah ? candy apple," in effect, what is the relation between Sarah and candy apple? She was given only one alternative, the new word "prefer," with which she replaced the interrogative marker, forming "Sarah prefer candy apple." Next, she was given the question, "Sarah ? apple candy," in effect, what is the relation between Sarah and apple-candy? Now the alternative given her was "not-prefer," the standard training device for the negative case, which amounts to the negative particle appended to the name for the positive case. She replaced the interrogative marker, forming the sentence, "Sarah no-prefer apple-candy." The names of two other foods for which preferences were known, viz., "banana" and "raisin," were substituted for "apple" and "candy," and she was given comparable training with them. She was then given the usual choice test, requiring her to decide between the use of "prefer" and "not-prefer," which she passed at the 80% level, and then a transfer test involving new cases for which preferences had been previously established, which she also passed at about the same level.

2.3.1. "Want" was taught in much the same way, though in this case we used deprivation for X and an essentially ad lib supply of Y to establish the difference between "Sarah want X" and "Sarah no-want Y." Subsequently, the mainte- nance conditions for X and Y were reversed. Sarah was then asked "Sarah ? X" and "Sarah ? Y," in effect, what is the relation between Sarah and X? and between Sarah and Y? Her alternatives in both cases were "want" and "no- want." Her ability to answer correctly at better than the 80% level established that her use of "want" was tied to and appropriately governed by specifically deprivation for the item in question.

2.3.2. One could argue that "want" does not mean deprivation in all cases; one might say he wants something even though he has not been deprived of it. In my opinion that is the meaning of "like" which I take to express a state of desire that is independent of the antecedents responsible for the desire. I see "want" as being more specific, as expressing a state of desire that is specific to the antece- dents of deprivation. But this discussion is irrelevant. In teaching the animal words that map her presumed internal conditions—e.g., states of desire that are and are not specific to the parameter of deprivation—we can arrange the distinc- tions in any way we like. We can seek a high correspondence with the vernacular, or we can ignore it and only fall into correspondence with it because the verna- cular inevitably reflects many, perhaps most, of the distinctions of which we are capable.

2.3.3. The last point of interest here is whether we should teach her psycho- logical state words at all. We could strand her in an operational format. Without psychological terms, she would have no alternative but to say, for example, "It's been a long time since I had bananas; If I were given bananas, I would eat them promptly and fast; the probability that I would eat bananas is high; etc." The last alternative is to introduce psychological terms, but not as we did, by manip- ulating conditions that presumably alter the states mapped by the words, rather by explicit definition. That is, for example, teach her "want" as the equivalent of some or all of the above descriptive statements.

3. There are contributions which the subject makes that owe little or nothing to the training program.

3.1. The abstractness of the definition that the subject gives to each language element is not modulated by the training program. For example, the inter- rogative marker, introduced in the context of *same-different* constructions, seemed nevertheless to stand for "missing element(s)" in other constructions. Indeed, though introduced originally in terms of a relation between actual ob- jects, it functioned later in relations between linguistic elements. Suppose we had found her definitions to be too specific. What could we have done to make them more abstract? In a word, nothing. The training did not modulate this important factor for the simple reason that we do not know how.

3.2. All the transfer tests she passed contribute to the same point. From the application of "same" and "different" to objects other than those used in training, through numerous intermediate cases, up to the most recent cases where the conditional particle was applied to atomic sentences different from those used in training, there was a common factor. She transferred the new particle to material different from that used in training.

3.2.1. Her most impressive or visible transfer was in the use of the quantifiers "all, none, one, some." She had been taught all of them with the use of sets of crackers that varied in terms of shape. Without additional training, she applied the quantifiers not to sets varying in shape, but to sets varying in color in one case and size in another.

Acknowledgments

This research was made possible by the patience and ingenuity of Mary Morgan, the main trainer, and of the others who assisted at various stages of the project, viz., J. Olson, Randy Funk, Deborah Peterson, Jon Scott, and Ann Premack. The research was supported by NIH grant MH-15616.

REFERENCES

Piaget, Jean. *The Language and Thought of the Child*, (3rd ed.) London: Routledge & Kegan Paul, 1962.
Premack, David. A functional analysis of language. *Journal of the Experimental Analysis of Behavior*, 1970, **14**, 107-125.
Premack, David. Language in chimpanzee? *Science*, 1971, **172**, 808-822.
Tolman, Edward C. *Purposive Behavior in Man and Animals.* New York, Century, 1932.
Vygotsky, L. S. *Thought and Language*, edited and translated by E. Hanfmann and G. Vakar, Cambridge, Mass.: M.I.T. Press, and New York: Wiley, 1962.

CHAPTER 4

The Habits and Concepts of Monkeys[1]

Donald R. Meyer

Monkeys learn concepts, and this affects the way that monkeys learn discrimination habits. Monkeys learn habits, and if they retain them, they fail to make use of their concepts. Thus, a trained monkey is a cognitive machine, but, like human cognitive machines, employs ideas only if a problem it must solve cannot be solved in any other manner.

These are strong statements, and the purpose of this paper is to show that they are also reasonable. The argument is based upon the outcomes of studies derived from the work of Harry Harlow. The concepts in question are termed learning sets, and processes of learning-set formation were first widely known about when Harlow (1949) presented his evidence that "monkeys learn to learn."

The roots of that evidence can now be traced back to approximately 1938, when Harlow and his students devised the apparatus which made his discoveries possible. At first glance, the Wisconsin General Test Apparatus (WGTA) does not seem like a great innovation. All that it amounts to is a cage for the monkey, a tray that carries stimulus objects, and screens that are used to permit the operator to set up problems between successive trials or to hide behind while trials are in progress (Fig. 1). The test tray commonly contains small wells in which bits of food can be concealed, and the monkey's task is to choose between objects which cover filled and unfilled foodwells. Most WGTA's, although not all, are powered by the operator's muscles; a few have motors that raise and lower screens, but these are the exceptions to the rule.

[1] The experiments described were supported in part by United States Public Health Service Research Grant MH-02035 and were conducted in part by fellows supported by United States Public Health Service Training Grant MH-06748.

ODD SHAPED OBJECT MOVED OFF FOOD WELL

FIG. 1. The Wisconsin General Test Apparatus, showing a monkey solving what is termed the Weigl oddity problem—to select, conditional upon a background cue, the "form-odd" or the "color-odd" object of a set of three objects.

Except for those who know it, the WGTA impresses most contemporary students of animal behavior as something that belongs on the shelf with the Hipp chronoscope. However, primatologists have spent many years of effort to develop other formats, and no one has yet devised another apparatus that has even come close to being as efficient for training monkeys as this old device (cf. Meyer, Treichler, & Meyer, 1965). The automatic systems which have thus far been proposed as successors to the WGTA are handsome apparatuses, but nonetheless are worthless for experiments concerned with learning sets.

As Harlow first recognized, the layout of the tray and the kinds of stimuli that one employs determine very largely how successful one will be in working with the WGTA. The most important rules are that small, solid objects are highly effective cues for monkeys, and that the animals must be required to touch the objects when they make their choice responses. Although much attention has been given to the question as to why these rules are as they are, the reasons behind them have yet to be discovered and must therefore be described as obscure. However, the effects produced by minor deviations from either rule are very large indeed, so large as to be almost unbelievable unless one has observed them for himself.

In all of the experiments to be considered here, the objects employed as stimuli were "multidimensional, common-use" objects, or as some workers term them, "junk" objects. Examples are tin cans, bathtub stoppers, ashtrays, and parts of children's toys. Objects of this kind do not lend themselves to automated training situations, and cannot be specified as readily, say, as a patch of red light of such-and-such a wavelength or a set of stripes of such-and-such widths. However, the questions we ask with such objects are not psychophysical questions, and manual presentation is a cheap price to pay for the genuine difference that their use has been shown to make when we teach monkeys concepts.

In the older studies which we shall discuss, the objects merely rested on the test tray. However, in experiments since 1961, we have made use of objects that are glued to flat plaques of ordinary thin masonite. These plaques, or bases, are then slipped into U-shaped holders that surround each of the test tray's food-wells. This is not a trivial detail. A mounted object and a free-standing object are equally discriminable for monkeys, but if and only if the animal is forced to touch the object rather than the base when it makes its first choice response. Stimulus-response contiguity effects have now been shown to be so important that unprotected bases ½ in. wide can result in a tripling of the number of trials that it takes to train monkeys to discriminate objects not displayed upon bases.

In all of the studies, the animals were trained on two-choice object discrimination problems. The common procedures were as follows. On any given trial, the monkey was faced with a choice between two stimulus objects. One of these objects, designated A, covered a foodwell in which a food reward was concealed. The other object, designated B, was positioned above an empty foodwell. The left-right positions of the A and B objects were governed by series which were balanced in a manner such that, in the long run, there was no correlation between where an object appeared upon the test tray and its probability of being correct. Hence the monkey's task was to discover which object covered the concealed food reward, and to choose that object irrespective of changes in its trial-by-trial positions on the test tray.

In most of the experiments, data were obtained from other training paradigms as well. However, in general, the other paradigms were also two-choice procedures in which one of the objects, and that object only, was reinforced on any given trial. Each of these procedures is simple to describe if we disregard positional considerations and then say only which object was correct on any given trial in a series. In all the latter paradigms, the A object was the first rewarded member of a pair, but the B object was subsequently rewarded for at least one trial within a series.

The subjects for all of the experiments were wild-born young macaque monkeys. Most of them were rhesus (*Macaca mulatta*), but a few were *M. speciosa* (or *M. arctoides*). As is now well known, these two kinds of monkeys have very different personalities (Kling & Orbach, 1963), but we have observed, in agreement with others (Gross, 1966; Schrier, 1966), that there are no important differences between the rhesus monkey and the speciosa monkey insofar as rates of formation of conventional object-discrimination learning concepts are concerned. Hence we shall not be overly concerned with what kind of monkeys

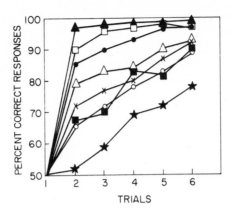

FIG. 2. Object-discrimination learning-set formation as originally described by Harlow (1949). Preliminary discriminations: (★) 1-8; (○) 9-16; (■) 17-24; (X) 25-32. Discriminations: (△) 1-100; (●) 101-200; (□) 201-256; (▲) 257-312.

were employed in each study, and we tentatively think that the basic outcomes now to be reported could be obtained with either animal.

Let us first consider the best-known result respective to learning sets in monkeys. Harlow (1949) obtained it by training monkeys on a series of six-trial, two-choice, common-use-object discrimination problems. Each successive problem was arranged with pairs of objects which the animals had never seen before; in the course of the study, the monkeys encountered several hundred such pairs of objects. For each of the problems, the rule was the same: A is reinforced and B is not. The monkey's task was thus to identify A, and thereafter choose only A.

The outcome of the study is shown in Fig. 2. In this graph, trial-by-trial group mean performances are plotted for successive large blocks of problems. These data demonstrate that monkeys given training on many discrimination problems eventually are able to master new problems of the same type in a single trial. Harlow, accordingly, believed that he had shown that monkeys are capable of insight, but also that insight develops gradually from a background of trial-and-error learning.

My own observations with respect to learning sets began with my doctoral research. In one of these projects (Meyer, 1951a), which was also concerned with drive as a variable in learning, I studied the formation of reversed-discrimination learning sets. The monkeys, before the main study was begun, had formed discrimination learning sets, and were next exposed to a series of problems in which, after 6 to 10 discrimination trials, the A object of a novel pair of objects became incorrect and the B object became correct for an additional 8 training trials.

Harlow (1950) had already explored this paradigm, and had found that the having of an object learning set does automatically equip a monkey to reverse discriminations as quickly as to learn discriminations. It was not clear why, if single trials sufficed to establish a discrimination in a single trial, single trials would not also serve to cancel out the effects of prereversal training. However, the results of Harlow's study showed that reversal sets, while learnable by monkeys, develop only after the animals are trained on a series of reversed discriminations (Fig. 3).

Harlow observed that the changes which occur when monkeys form reversal-learning sets are highly reminiscent of the changes which occur when monkeys form discrimination sets. Thus, if one plots the intraproblem course of learning following the first reversal trial, there develops what appears to be a sharp discontinuity between trials 1 and 2 of the function. I found the same thing in my experiment, and then noticed something interesting, namely, that a single, simple asymptotic curve could be passed quite successfully through all the data points for trials 2-8, regardless of how well the animals were doing on trial 2 (Fig. 4).

I took this to mean that "learning to learn," at least of reversed discriminations, obscures but does not alter the processes involved in learning by a "set-naive" monkey. To test this idea, I subsequently gave my animals reversal-cue problems, a paradigm designed to countercondition or extinguish the reversal learning set (Meyer, 1951b). Thus it seemed to me that if habit mechanisms are

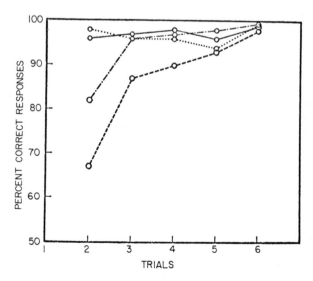

FIG. 3. Formation of the object-discrimination-reversal learning set as originally described by Harlow (1950). Problems: (- - -) 1-28; (•—→) 29-56; (—) 57-84; (• • •) 85-112.

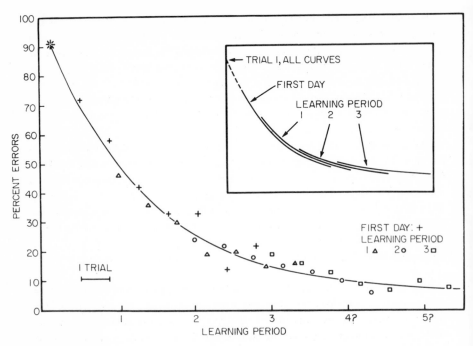

FIG. 4. Replication and analysis of object-discrimination-reversal learning-set formation by Meyer (1951a).

not changed by learning-set formation, it should then be possible to show that monkeys which have unlearned a rule will approach a problem in the same general way that they approached that problem before they learned the rule in the first place.

Reversal-cue problems, like reversal problems, begin as discrimination problems. They then have one trial on which the A object is incorrect and B is correct. Then they are completed by a series of trials on which the A object is correct once again, and hence if the monkey reverses on cue, it commits a second error after the rereversal of the cue-reward contingencies. To solve such problems, the monkey, in effect, must learn to disregard reversal trials instead of, as before, responding to these trials as cues to abandon A objects.

As Fig. 5 shows, the monkeys responded to the first group of these problems as expected: They made many errors on reversal-cue trials, and then made many errors on the first trials on which the initial discrimination rules were reinstated. However, the animals then quickly returned to the strategy of choosing A objects, and after a series of reversal-cue problems, learned to ignore reversal trials.

At this point, I retrained the monkeys on a series of "genuine" reversed discriminations, and this gave the intraproblem learning function which is reproduced in Fig. 6 and therein compared with the animals' initial performances on problems of this kind. The two curves differ, but not by very much, and their similarities impressed me because a year in which the animals had had varied training on other kinds of problems had intervened between the studies in which they had been generated. Accordingly, it seemed that extinction of the set had served to return the animals to much the same condition they had been in before the set was established in the first place.

With these facts as background, I now wish to add some new and unpublished observations which bear upon these ancient suggestions that a monkey can learn and unlearn a simple concept without, in the process, being changed very much

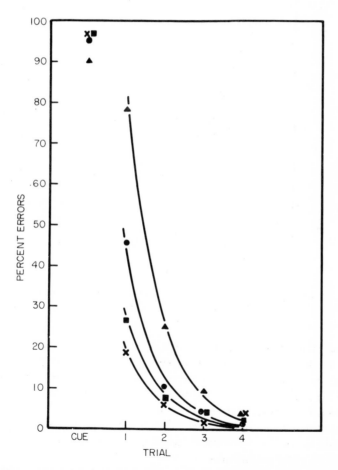

FIG. 5. Course of countertraining or extinction of the reversal-learning set via presentations of reversal-cue problems as described by Meyer (1951b). Two-week periods: (▲) 1; (●) 2; (■) 3; (X) 4.

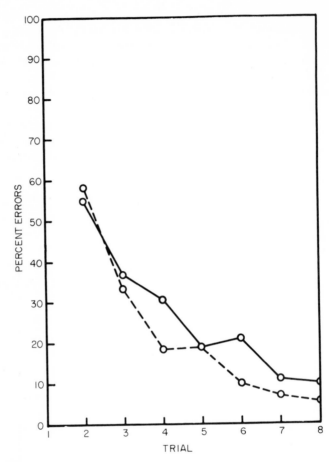

FIG. 6. Comparison of initial and postextinction object-discrimination-reversal learning (Meyer, 1951b). (—) Original reversal learning; (- - -) relearning, 1 year later.

with respect to the way it learns a habit. The first of these studies was carried out by Mildred H. Lopopolo. It involved the cyclical extinction of reversal-learning sets. A group of five monkeys was alternately trained on reversal and reversal-cue problems until the reversal set had been formed seven times and had been countertrained seven times. The question was whether training of this kind distorts intraproblem learning functions, or whether, instead, such functions have a rate which is wholly independent of the momentary strength of the set as measured by trial-2 performance.

The problems were presented on a test tray with three foodwells, and all three positions were employed. On the first trial of each problem, the two objects to be discriminated were placed above the two outside foodwells. Afterward, on each successive trial, the objects were presented above the two foodwells that had not contained the food reinforcement for the previous trial. This procedure was followed as a means of suppressing what Harlow (1959) has described as the differential-cue error factor and of minimizing other positionally related strategies or "hypotheses" (Levine, 1965). In other respects, the general training methods were the same as have already been described. Only one choice was allowed on each trial and only one object was rewarded, and the relative left-right positions of the objects were changed in counterbalanced orders. The foods which were used for the reinforcements varied with the monkeys' preferences, and generally were peanuts, small bits of apple, raisins, or candied cereals.

The animals had previously been trained on a series of six-trial object-discrimination problems. For the study here considered, they were first reminded of the rules of object-discrimination learning by being given object-discrimination problems which were each arranged with novel pairs of objects and were each 10 trials in length. Five such problems were presented daily, and training was continued with each monkey until it performed at the 85%-correct-response level on the second trials of the problems. This criterion was met when the monkey had, within a series of 20 problems, performed correctly on the second trial of any 17 of the problems.

The main experiment consisted of seven cycles of training in reversal-learning sets and training on reversal-cue problems. Each reversal problem began as a conventional object-discrimination problem and, as in the studies described above, was each arranged with novel object pairs. Reversals occurred on trials 5, 6, 7, 8, or 9 of a problem, and the initially negative B object was reinforced for four total trials. The variations were included to keep the subjects from forming nth-trial-reversal sets, a skill that is well within the repertoire of most sophisticated rhesus monkeys. Each reversal problem thus had a length of from 8 to 12 total training trials, but, since the monkeys had formed object-learning sets, prereversal levels of performance after 4 discrimination-learning trials were virtually as high as they were after 8 discrimination-learning trials. Five reversal problems were given each day, with the order of presentation of problems of different total lengths being balanced by reference to a 5 × 5 Latin square.

Each reversal phase for each training cycle continued for each monkey until its performance on the first postreversal trial was 85% correct (17 correct responses in any block of 20 successive problems). Reversal-cue training was then introduced, and the technique was exactly as in reversal training except that the object-reward contingencies were reversed for one trial, and one trial only. This reversal trial occurred on either the 5th, 6th, 7th, 8th, or 9th trial of a problem, and was then followed by a reinstatement of the prereversal cue-reward contingencies for three further trials. Reversal-cue training continued until the animal's performance on the first trial following the one trial on which the B object was rewarded reached 85% correct.

My prediction was that these acquisitions and extinctions of the reversal learning set would have no effect whatsoever upon the intraproblem rates of learning of either reversal or reversal-cue problems. That is, I expected that an intraproblem curve based upon data collected at the beginning of each cycle of set-acquisition or set-extinction would comfortably fit the intraproblem data obtained at any point within the cycle. According to my theory, the sole effect of formation or extinction of the reversal set would be to change the level of performance on the trial immediately after the reversal-cue trial. With that point established, I expected that the rest of intraproblem learning would proceed at the same rate that intraproblem learning proceeded before the monkeys first formed the reversal set.

Our strategy in testing this idea was to begin by plotting mean functions for intraproblem learning on the basis of performances obtained on first days of training on reversal and reversal-cue problems. Next, we had to choose where to look at such functions as affected by learning-set formation, and the latter decision was complicated by the fact that successive acquisitions and extinctions of the reversal learning set did not always take the same amount of practice to complete. We resolved this problem by assuming that the progress of learning-set formation or extinction within cycles takes place at a constant rate. We then generated intraproblem functions during set formation and set extinction by pooling the data for the days which were halfway between the first day and the criterion day for each cycle.

Figure 7 shows the results. Each of these functions is derived from the data from five monkeys trained on seven different occasions, and hence each point is based upon a total of 35 (or 5 X 7) training trials. To us, it seemed that the predictions were fulfilled in that the middle-day intraproblem functions following the trial-2 discontinuities are very nicely estimated by the intraproblem functions obtained from the first-day data. Further, the reversal-cue functions (extinction) were steeper than reversal-learning functions, as would be expected from my old explorations of reversal and reversal-cue learning.

To me, the main point of this result is not the fact that it fit my expectations so exactly. Indeed, when we consider the coarseness of the data, we need to be conservative about them. However, what is clear, and I believe, of great importance is that habit learning is not, as Hebb (1949) described it, "distorted out of all recognition" by learning-set formation. Instead, in the language of Restle's (1958) set model, extinction seems to "unadapt" the cues which Restle visualized as undergoing "adaptation" during acquisition of the set. I view this conclusion as being grounds for cheer for those who find habits interesting; at least, it suggests that they are dealing with events which are governed by stable mechanisms.

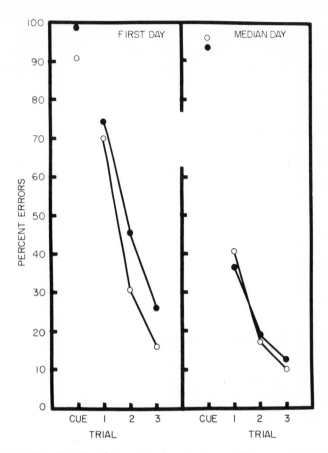

FIG. 7. Intraproblem learning of reversal and reversal-cue problems as measured by Lopo-polo during successive formation (●) and extinction (○) of the object-reversal learning set.

I am not so sure that the same thing is true of mechanisms underlying sets. At least, we have found that cyclical extinction of reversal sets results in large changes both in the rate at which the set is formed and the rate at which the set is extinguished. Figure 8 traces the progress of the monkeys in Lopopolo's study over all seven cycles of training in terms of how many mean problems the animals required to reach the criterion of 85% correct trial-2 responses. The scores at all points exclude the 20 problems during which criterion was met; thus, for example, the monkeys took a mean of 50 problems of reversal-problem practice to form the first concept in the series.

The shape of this function will be of some interest to students of subprimate species because it reminds us of what happens when a rat is trained on successive

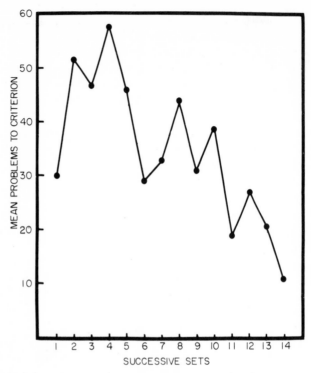

FIG. 8. Rates of formation and extinction of the reversal-learning set as measured by Lopopolo. Odd-numbered points are for formation; even-numbered points are for extinction.

reversals of a single simple left-right position habit (e.g., Dufort, Guttman, & Kimble, 1954; see also Warren, 1965). It tempts us to think that the learning, by monkeys, of "what one reversed discrimination trial means" is closely akin to the learning, by rats, of whatever it is that permits a trained rat to give up a left-turning habit for a right-turning habit after one nonreinforcement. However, the problem with such an argument is that rats can learn a one-trial reversal, but are barely able, after thousands of problems, to do much better on a new example of even a simple discrimination problem than they did on the first such example they encountered.

I think that the source of at least a large share of the species differences we see in concept learning is a species gradation with respect to the effects of changes in stimuli upon disinhibition of set-suppressed error factors. Such a point of view is consistent with the fact that the only formal test situations in which subprimate animals behave in ways which clearly indicate that they have learned rules or concepts are test situations which are stable insofar as the stimuli involved are concerned. Object- and positional-alternation problems are illustrations of such situations, as are the particular varieties of oddity and matching-from-sample tasks that rats, or even birds, come to master.

I think that what happens when monkeys that have formed a reversal learning set are given countertraining is reasonably similar to what occurs in rats when new stimuli are introduced. I also believe that the effects of countertraining are not unlike, but rather are merely more profound than the effects of introductions of new kinds of rules into the games we give the animals. These two ideas are closely related to Harlow's (1959) concept of regression, that is, his notion that unsolvable problems induce progressively more primitive approaches to the solution of the problems.

The fact that reversal learning following extinction resembles preset-reversal learning is, in my opinion, the clearest evidence we have for the concept of regression. However, the resemblance that has thus far been established is between group-mean intraproblem functions, and error-factor theory, in which we find the concept, considers mean functions to be net consequences of many underlying processes. We thus have a problem of understanding how the mean curves can end up being simple when their points are specified by the results of many choices by a number of monkeys of varying talents whom we know to be learning several different things at several different rates.

I suspect from the fact that mean intraproblem functions are among the most stable of the learning curves known to comparative behaviorists that some kind of process serves to establish a relationship between the several processes which error-factor studies have revealed. In other words, I think that Harlow's error factors and the closely related "hypotheses" defined by Levine (1965) in his model of learning-set formation are not independent in the sense of being manipulable without producing changes in remaining error factors or hypotheses except for enhancement or reductions in their strengths in proportion to the changed share of choices which the manipulations in question would leave to the remaining error factors or hypotheses. Instead, I believe that extinction of a set, when it reinstates one error factor, reinstates others in degrees predictable from the relative strengths of the error factors in a given group of naive animals.

I think that this notion can best be conveyed by the statement that a "set-naive" monkey already has a set or a strategy of which several error factors are measures. These are the stimulus-related error factors which change most when other sets are formed, including the differential-cue error factor and object and position preferences. The fact that these factors are stimulus-related suggests that, to a first approximation, the "set-naive" monkey is controlled by a concept "that stimuli which this situation shares with others have something to do with which responses are productive of rewards."

I view this idea as consistent with the outcomes of the second of our new experiments. This work was done in association with Drs. Lewis Bettinger, Roy

A. Anderson, David A. Yutzey, and David A. Dalby. The study was carried out with 11 monkeys that had first formed object discrimination sets, and dealt with the effects of retentions of habits upon the utilizations of these sets. In general, the experiment involved presentations of a single pair of recurrent cue objects which were interspersed between practices on varying numbers of conventional six-trial problems. The recurrent cue objects were a red circle and a green triangle which were presented as low, planometric (i.e., flat) objects cemented to the faces of wedges whose surfaces were tilted upward toward the monkeys. The wedges, in turn, were attached to thin plaques which served as foodwell covers and could be inserted into U-shaped guards of the type previously described.

The training methods were developed from two pilot studies. The general paradigm was as follows. First, the animal was trained to choose, say, the red circle until that stimulus had been selected on five successive training trials. Then the monkey was given training with 0, 1, 2, or 3 novel-object problems, each for a total of six trials. Then the animal was trained to choose the red circle or the green triangle until, once again, it made five correct responses in a row to whichever of the latter two recurrent objects was then designated as correct. The latter transfer test also set the stage for further interpolated novel-object training and a further recurrent-object transfer test. The recurrent circle-triangle problems and the conventional interpolated problems were arranged in a series so that, over 28 days of training, six "positive" and six "negative" transfer tests were presented to each monkey after 3, 2, 1, and 0 interpolations of novel-object problems.

The circle and the triangle were equally often reinforced. A positive-transfer test was one in which the cue-reward contingencies that happened to prevail before interpolation of novel-object problems remained the same following such training. A negative-transfer test was one in which the cue-reward contingencies of the recurrent problem reversed following interpolation of novel-object problems. A positive transfer test with no interpolation was thus merely a continuation of practice with the circle-triangle problem until the same stimulus had been responded to five times in succession and then five further times in succession. Conversely, a negative-transfer test with no interpolation was one in which the monkey was trained to choose one familiar cue until it had done so five times in succession, and then to choose the other familiar cue until it had done so five times in succession.

The presentations of recurrent problems had no obvious effect upon the monkeys' levels of performance with the interpolated novel-object problems. Their mean trial-2 levels of performance on these problems was roughly 85%, which was within 2% of the mean trial-2 value for the group during the last phase of their preliminary learning-set training. The group mean performance on the sixth trials of these problems was 95% correct responses, and hence each could be thought of as having been "fully learned" before it was succeeded by the next task in the series.

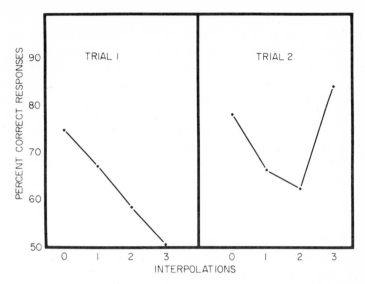

FIG. 9. Learning-set suppression produced by retention of stimulus-specific habits. Data from the experimental program of Bettinger, Anderson, Yutzey, Dalby, and Meyer.

Figure 9, left, gives the same animals' performances on first trials of re-presentations of the circle-triangle recurrent problem as a function of number of interpolated problems. These are the outcomes obtained from situations which should yield positive transfer; unfortunately, negative-transfer situations did not yield interpretable results. The positive-transfer data show that three interpolations of novel-object problems reduced first-trial choices to chance; that is, the monkeys progressively forgot which cue, the circle or the triangle, had been correct on the last occasion of the presentation of this pair of objects.

Figure 9, right, is a resume of group-mean trial-2 performances of the circle-triangle problem, again as function of number of interpolated novel-object problems. Reasonably enough, these levels were usually higher than they were on trial 1, but the function is not exactly what would be expected if one were to think that it always helps to remember what has worked in the past. Thus the performances on trial 2 are best after zero and three interpolations, and the performance after three interpolations is roughly the same as it was on second trials of the novel-object problems which were used as interpolated tasks.

I take this to mean that habits, if retained, suppress discrimination learning sets, and generally that habits, like extinction of sets, induce conceptual regressions. That is, I believe that the sharp increase in trial-2 performance which occurs after three interpolations reflects the monkeys' treating the recurrent problems as problems they have never seen before. They then apply a rule which we know that they know from the way they solve novel-object problems, but seemingly cannot employ so long as they retain at least a little of their last habit. I think that this happens because retained habits induce the set of "set-naive" monkeys, and also that this gives us an explanation for the fact that intra-problem functions after second trials follow their characteristic courses.

We have, in this finding, an effect which seems to be a converse of Riopelle's (1953) supposed demonstration that learning-set formation is accompanied by suppression of retention of specific habits. Riopelle's experiment involved train-ing monkeys on a series of discrimination problems in which some problems were reversals of problems that the monkeys had previously encountered. Rio-pelle found that, initially, reversals produced substantial negative transfer, but that the amount of this transfer was reduced as monkeys formed discrimination sets.

Figure 10 is a new illustration, from the studies of Bettinger *et al.* of transfer suppression as measured by a slightly different methodological approach. The technique involved the training of a group of five *M. speciosa* monkeys on a series of 392 six-trial "common-use-object" discrimination problems. The first 98 were novel problems, but the second 98 were the first 98 with cue-reward contingencies reversed. The third 98 were the first 98 with the original contin-

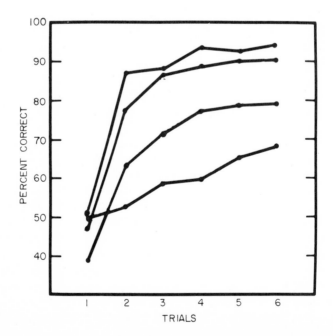

FIG. 10. Apparent development of suppression of stimulus-specific habits as a function of discrimination-learning set development. Data from the experimental program of Bettinger, Anderson, Yutzey, Dalby, and Meyer.

gencies restored, and the fourth 98 were the first 98 with cue-reward contingencies reversed once again.

The first-block data are highly reminiscent of what one observes with rhesus monkeys. That is, the function starts at 50%, and then tends to rise very slowly. The second-block curve, on the other hand, begins at approximately 39%, which means that the animals were able to remember a fair amount about which objects were rewarded when they last saw them far back in the series. However, in later blocks, the first-trial performance approaches and then arrives at chance, and hence behaves exactly as predicted from the concept that learning sets suppress specific transfer.

However, one can argue that a find of this kind can be interpreted in other ways. Thus, one could say that in the later training phases, the monkeys treat old objects as neutral objects because they have been both A and B objects at least once before there is any sign whatever of reduction of specific-habit transfer. Also, recent studies of Stollnitz and Schrier (1968) and of Schrier (1969) have raised some doubts about there being any necessary close relationship between set formation and suppression of specific-habit transfer. One important argument for this last conclusion is the fact that highly set-trained animals do not show suppression of specific-habit transfer until they have been trained for a number of weeks with probe problems of the type that Riopelle (1953) employed to measure the occurrence of suppression. Stollnitz and Schrier (1968), accordingly, suggest that Riopelle's phenomenon is not a set-induced suppression, but a learning by monkeys to reverse discriminations when they recognize objects as familiar.

I now wish to add some new information which bears upon this kind of interaction which presently is still in the process of being collected by Jacqueline Conner in our laboratory. Conner's group consists of 10 rhesus monkeys that underwent a long taming process before being trained on a preliminary group of 12 six-trial discrimination problems. We plan, in the future, to see if the individual strategies the monkeys used in solving these problems as naive animals can be recovered via set-extinction procedures. To do this, of course, we must set-train the monkeys, and to get them set-trained is a major purpose of the study that is currently in progress. But, since we believe that specific-habit learning is a function of stable mechanisms, we have felt very free to engage in deviations from the usual set-producing training methods.

Conner's first inquiry has been directed toward the question as to whether suppression develops in the absence of training on reversed discriminations. Her study thus far has involved the presentation of a series of six-trial discrimination problems. On days 1 through 6 of each successive block, six novel problems are presented. On days 7 through 10 of each block, three problems from days 1 through 4 are re-presented in alternation with three novel problems so that measures of retention of the old discriminations can be had after intervening practice with a total of 36 other six-trial discrimination problems. When the old problems are represented, their cue-reward contingencies are always the same as they were when the problems were first practiced, and hence the monkeys never encounter the reversals which Stollnitz and Schrier (1968) have suggested to be crucial for transfer suppression to occur.

There are five training sessions per week, and hence a block of problems is completed in 2 weeks. After a preliminary 2-week block which served to introduce the monkeys to the schedule (and corresponded roughly to the first, or "flat" stage of formation of discrimination learning sets), the animals were entered upon a training schedule in which they are given three blocks of training over 6 weeks, are then rested for 2 weeks, are then given three blocks of training over 6 weeks, are then rested for 2 weeks, and so on.

Figure 11 shows what Conner has observed in the first nine blocks of her study. The three functions summarize trial-1 and trial-2 performances on "old" discrimination problems and, as an index of set formation, trial-2 performances on "new" discrimination problems. Each point is based upon the six old and six new problems presented on the seventh and eighth days of each block because the phenomena of interest are highly transitory. Hence, each data point in each of these curves summarizes 60 choice responses, or 6 choice responses by each of the 10 rhesus monkeys in the group as a whole.

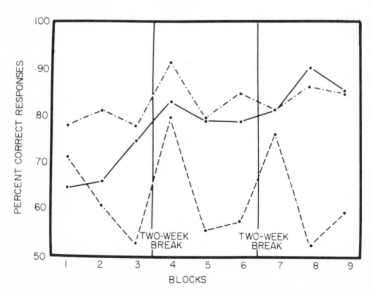

FIG. 11. Lack of correspondence of course of development of the object-discrimination learning set (trial-2 novel-object function) and development of transfer suppression (trial-1 repeated-object function). (——) Trial 2 novel problems: (- - -) trial 1 repeated problems; (•—•) trial 2 repeated problems. Data from the experiment of Conner.

The functions show, first, that transfer suppression can develop without reversal training. The measure of suppression is trial-1 performance on "old," or repeated discrimination problems, and trial-1 performance on these problems declines from block 1 to block 2 to block 3. Hence, although training on reversed discriminations may serve to suppress specific habits, Conner has shown that Stollnitz' and Schrier's theory that suppression necessarily depends upon reversal training is wrong.

I must hasten to add that this same group of curves is equally contrary to the view that transfer suppression is fundamental to the process of learning-set formation. Consider the effects of breaks in training. Insofar as trial-2 levels of performance on novel-object problems are concerned, a 2-week interval between successive blocks has almost no effect whatsoever. However, a rest of the same length disinhibits suppression of retention of old habits, at least insofar as such retention is measurable in terms of trial-1 performances. Accordingly, the Stollnitz-Schrier criticism of Riopelle's theory has its merits, and the present outcomes are inconsistent only with their theory of the source of the suppression.

My own present thoughts are that transfer suppression results from retroactive interference, and hence is a problem for theory of habits rather than for theory of concepts. I view its relationship to concept formation by monkeys as being largely indirect. Thus, if retentions of habits yield suppression of a monkey's use of learning sets, it seems necessary to believe that procedures which enhance such retentions must serve to retard the rates at which learning sets are formed. Some of Conner's findings support this conclusion, but fall somewhat short of proving it, and hence I regard it as a point to be settled by investigations of the future.

Finally, I should like to summarize these findings in the form of three general principles. The first is that habits interfere with utilizations of concepts by infrahuman primates; it is also probably true that old habits interfere with the formation of new concepts. The second is that acquisition of a concept is not a special kind of habit learning, but rather can occur without affecting mechanisms of stimulus-specific acquisitions. The third is that stimulus-related error factors reflect a common tendency in monkeys, and this, instead of independent tendencies, becomes suppressed when simple learning sets are formed.

I think that these rules apply to human beings as well as to infrahuman primates. Consider, for example, the tendency of men to become more rigid in the ways of doing things as a function of increasing age. Our data suggest that this process is due, in part at least, to our accumulating habits which not only prevent us from using old ideas but also interfere with our learning of new conceptual rules. I am sufficiently convinced that this is so that, as a man whose business it is to try to find relationships between sets of facts, I make it a point not to use lecture notes when I meet my university classes. These classes do not go as smoothly as they did when I had all my stimuli before me, but my present impression is that things come out of them which even *I* find new and interesting.

REFERENCES

Dufort, R. H., Guttman, N., & Kimble, G. A. One-trial discrimination reversal in the white rat. *Journal of Comparative and Physiological Psychology*, 1954, **47**, 248-249.

Gross, C. G. Learning set: Comparison of *Macaca mulatta* and *M. speciosa*. *Psychological Review*, 1966, **18**, 529-530.

Harlow, H. F. The formation of learning sets. *Psychological Review*, 1949, **56**, 51-65.

Harlow, H. F. Performance of catarrhine monkeys on a series of discrimination reversal problems. *Journal of Comparative and Physiological Psychology*, 1950, **43**, 231-239.

Harlow, H. F. Learning set and error factor theory. In S. Koch (Ed.), *Psychology: A study of a science*. Vol. II. New York: McGraw-Hill, 1959.

Hebb, D. O. *The organization of behavior*. New York: Wiley, 1949.

Kling, A., & Orbach, J. The stump-tailed macaque: A promising laboratory primate. *Science*, 1963, **139**, 45-46.

Levine, M. Hypothesis behavior. In A. M. Schrier, H. F. Harlow, & F. Stollnitz (Eds.), *Behavior of nonhuman primates: Modern research trends*. Vol. I. New York: Academic Press, 1965.

Meyer, D. R. Food deprivation and discrimination reversal learning by monkeys. *Journal of Experimental Psychology*, 1951, **41**, 10-16. (a)

Meyer, D. R. Intraproblem-interproblem relationships in learning by monkeys. *Journal of Comparative and Physiological Psychology*, 1951, **44**, 162-167. (b)

Meyer, D. R., Treichler, F. R., & Meyer, P. M. Discrete-trial techniques and stimulus variables. In A. M. Schrier, H. F. Harlow, & F. Stollnitz (Eds.), *Behavior of nonhuman primates: Modern research trends*. Vol. I. New York: Academic Press, 1965.

Restle, F. Towards a quantitative description of learning set data. *Psychological Review*, 1958, **65**, 77.

Riopelle, A. J. Transfer suppression and learning sets. *Journal of Comparative and Physiological Psychology*, 1953, **46**, 2, 108-114.

Schrier, A. M. Learning-set formation by three species of macaque monkeys. *Journal of Comparative and Physiological Psychology*, 1966, **61**, 490-492.

Schrier, A. M. Learning set without transfer suppression: a replication and extension. *Psychonomic Science*, 1969, **16**, 263-264.

Stollnitz, F., & Schrier, A. M. Learning set without transfer suppression. *Journal of Comparative and Physiological Psychology*, 1968, **66**, 3, 780-783.

Warren, J. M. Primate learning in comparative perspective. In A. M. Schrier, H. F. Harlow, & F. Stollnitz (Eds.), *Behavior of nonhuman primates: Modern research trends*. Vol. I. New York: Academic Press, 1965.

CHAPTER 5

The Effects of Deprived and Enriched Rearing Conditions on Later Learning: A Review[1]

John P. Gluck and Harry F. Harlow

The fact that experiences of an organism early in life exert a pervading influence on its behavior later in life is often given the status of a truism. Actually, there now exists a wealth of experimental studies appraising the effects of various early environments, ranging from extremely deprived to greatly enriched, on the subsequent learning and intellectual abilities of animals. Generalizations drawn from this literature have led to the position that deprivation rearing attenuates all learning capacity, while rearing in enriched environments facilitates all learning. Until recently, discussions of these deprivation and enrichment positions were limited to the debate of developmental theorists. However, the initiation of a federally supported program (the Head Start program) designed to enrich the lives of some children exemplifies the extent to which these generalizations concerning the effect of enriched early experience have penetrated our thinking. Given this situation it appears relevant to examine the literature that has provided the medium from which these concepts have developed. It might seem unfair to uncover old skeletons at this time, but many procedural premises have been established on rather shaky research foundations whose objectives or experimental sophistication have not been adequately examined.

The review is limited to two general types of early experience: enrichment or supernormal stimulation, and deprivation or subnormal stimulation. The methodological procedures involved in creating such environments vary with respect to particulars from species to species. However, the general procedures may be characterized by the following:

[1] This research was supported by Grant FR-0167 and Grant MH-11894 from the National Institutes of Health to the University of Wisconsin Regional Primate Research Center and Department of Psychology Primate Laboratory, respectively.

Enriched Environment: Experimental subjects are reared with more than one other age-mate, together with extra stimuli such as geometric objects varying in size, shape, and color. Climbing and activity apparatus is also frequently available. The major exception involves studies of dogs, where enrichment is often produced, at least in theory, by making the experimental subjects pets of laboratory personnel.

Deprived Environment: The subjects are reared in an experimental enclosure smaller than that utilized in the enriched condition, devoid of either age-mates or extra stimulus objects. Normal rearing involves living with a smaller number of age-mates than in the enriched condition, in a smaller enclosure devoid of extra stimulus objects. The major exception here involves primate studies where subjects are generally not permitted actual physical contact with age-mates, but are permitted to see and hear other monkeys (called partial isolation). Further, in rat and dog studies, the subjects are rarely placed in the rearing conditions prior to weaning from the natural mother. Studies utilizing primates usually place subjects in rearing conditions within a day or two following birth.

Before proceeding directly to the evaluation of deprivation and enriched early experiences and later learning capacity, it is essential to note the behavioral effects of such experiences on activities other than learning. Rhesus monkeys reared in restricted environments exhibit deficits in exploratory, affiliative, communicative, emotional, sexual, and maternal behaviors (Harlow & Harlow, 1965; Sackett, 1965; Miller, Caul, & Mirsky, 1967). Partial isolation rearing produces quantitative deficits compared to rearing with mothers and/or age-mates. Rhesus monkeys reared in total isolation are socially devastated, appearing generally inactive, unresponsive to most stimulus changes, exhibiting almost no positive social behavior, and displaying high amounts of stereotyped and self-directed behaviors.

Rats reared in isolation have been reported to be hyperactive when tested in an open-field apparatus (Woods, Fiske, & Ruckelshaus, 1961). Sexual behavior of rats has been described as disrupted following a period of social isolation commencing 2 weeks after birth (Gerall, Ward, & Gerall, 1967).

Melzack and Burns (1965) reported that two behavioral features differentiate restriction reared dogs from normally reared littermates. "First, restricted dogs exhibit an extremely high level of behavioral arousal or excitement. Almost anything new in their environment, such as a slight modification of their home cage, is able to raise the level of excited activity. Second, restricted dogs have difficulty in attending selectively to environmental stimuli. They often dash from one object to another in a room, but rarely show sustained attention to any single object."

EXPLANATIONS OF EARLY DEPRIVATION EFFECTS

With regard to the above effects attributed to early deprivation experience, four general explanatory views are common.

Atrophy

Riesen (1966) demonstrated that deprivation of sensory input (dark rearing) can result in specific degeneration in sensory systems, particularly the visual system. As a result of deprivation rearing an animal may thus be relieved of specific neural apparatus serving as the substrate for a particular set of behaviors.

Developmental Failure and Potentiation

A related view concerns neural systems that are relatively undeveloped at birth and require postnatal stimulation in order to achieve full maturation. In other words, it is hypothesized that the external environment must contribute critical stimuli to complete neural maturation. If an organism is reared in an environment devoid of these critical stimuli, later behavioral deficits may emerge as a result of the action of an immature physiological system. Thus it is conceivable that enriched rearing may permanently facilitate the quality and quantity of behaviors related to maturationally potentiated physiological systems. Evidence for such an effect has been provided in rodents by the work of Rosenzweig, Krech, Bennet, and their coworkers. Their data suggest that active commerce with peers and other environmental objects account for a large portion of the increments in rats' brain growth and chemical activity produced by enrichment procedures (Rosenzweig, 1966).

Learning Deficits

Other explanations have postulated that deficits are caused by the failure of the rearing environment to provide experiences critical for learning basic perceptual-motor, emotional, social, and information-processing responses. Thus failure to learn to organize perceptual and motor responses early in life may permanently lower adaptive capabilities (Hebb, 1949). A similar interpretation suggests that the reinforcement contingencies operating during social interaction with species members are not learned because of deprivation rearing, thus leaving the animal incapable of responding appropriately to stimuli within a social context (Harlow & Harlow, 1965; Scott, 1962). Further, Sackett (1970) suggests that deprivation rearing blocks the learning of a more general type of response strategy, i.e., a failure to develop inhibitory control over responses that compete with behaviors demanded by changing environmental conditions.

Emergence Trauma

A final type of explanation assumes that behavioral deficits following rearing are a function of discrepancies in novelty, complexity, or intensity between rearing and postrearing stimulation (Fuller, 1967; Sackett, 1965). When these discrepancies are large they may produce, upon emergence from the deprivation-rearing condition, high levels of arousal or emotionality. These conditions then yield disoriented, hyperactive behavior in rats, or immobile, avoidance-type behaviors in monkeys. Early experience-produced deficits are thus viewed as emotional or attentional anomalies, produced by the interaction of rearing and postrearing stimulus input levels.

Given the impressive list of behavioral peculiarities found in deprivation-rearing studies and following from any of the common types of rearing effect explanations, it seems plausible to anticipate altered learning capacity in deprived subjects. Let us examine data appropriate to this generalization.

EXPERIMENTAL INJUNCTIONS FOR THE STUDY OF DEPRIVATION EFFECTS

The study of early experience effects on learning is beset by many methodological difficulties. Since it is widely assumed that the concept of learning capacity refers to fundamental behavioral processes, it is crucial that learning differences observed between differentially reared animals are not directly attributable to mere procedural artifacts. Toward this end, we present seven injunctions that we feel must be applied to the design of relevant experiments. Though we believe that all seven are extremely important, only the most pertinent departures from these guidelines will be emphasized in the body of the literature review which follows.

1. All experimental animals must achieve equal adaptation to the experimental situation or situations prior to the start of testing. Apparatus adaptation is difficult and a minimal number of basic learning situations is desirable.
2. Tests of learning which involve the transfer to the test situations of specific response relationships learned in the rearing situations are not acceptable.
3. The use of aversive stimulation such as electric shock is absolutely contraindicated since animals tend to be hyperemotional after confinement to deprived environments.

4. If the experimental animals are tested prior to adulthood, maturational data regarding normative test performance must be known. Otherwise, it will not be clear whether observed differences indicate a transient developmental slowing or a permanent adult deficit.

5. Generalizations regarding overall learning capacity should be based on the results of a comprehensive battery of tests which measures many dimensions of learning throughout the various maturational stages of the organism as opposed to an individual test or a few tests at a single point in time.

6. The performances of the individual subjects within each treatment should be reviewed when broad within-group variability is apparent to the experimenter. Such a review would reveal whether the treatment effects were pertinent in individual animals or groups.

7. The levels of emotional behaviors that might compete with learning performance must be identified and measured.

EFFECTS OF EARLY EXPERIENCE UPON LEARNING IN RODENTS

Much of this work was stimulated by Hebb's (1949) report that rats reared as free roving pets housed in his own home were superior problem solvers in comparison with rats reared in standard laboratory cages. Problem solving ability was assessed in a series of detour problems presented in the Hebb-Williams maze (Hebb & Williams, 1946). The Hebb-Williams maze consists of a start and goal box that remain in fixed positions separated by an open field in which barriers are placed in various configurations in order to obstruct direct passage from the start to the goal box. The more direct the route the better the score. Hebb conjectured that the richer experiences of the pet group permitted the development of cortical cell assemblies which made the home-reared subjects more able to profit by new experiences at maturity. However, no such facilitation was observed when the groups were run in a T maze. Since the pet-reared group certainly had encountered many open fields in Hebb's home compared to the cage-reared group, it appears difficult to reject the possibility of an adaptation bias favoring the pet-reared group.

Because Hebb utilized small groups without precise control of early experience, Hymovitch (1952) replicated the studies but employed larger numbers of rats under more carefully controlled rearing conditions. Rats were raised in an enriched environment or individually stuffed in stove pipe enclosures, or housed in solid metal structures containing activity wheels. The animals from the enriched environment were exposed to blind alleys, inclined runways, and small enclosed areas and obstacles. Subsequently the subjects were run on a series of Hebb-Williams detour problems. It appears that the selection of early experiences to which the enriched animals were exposed guaranteed their differential adaptation and, thus, superior performance. Though the enriched animals demonstrated facilitated Hebb-Williams performance, they failed to surpass the restricted subjects' performance when tested on a ten-choice point maze.

Continuing in the perceptual learning tradition, Forgays and Forgays (1952) attempted to isolate those elements present in the enriched environment responsible for the apparent facilitative effect. Rats were weaned at 26 days of age, then placed in one of seven experimental rearing conditions: (1) free environment with other animals and "playthings"—the playthings were not described in the written report, but from the single diagram offered those "things" included most obviously the barriers resembling those of a simplified Hebb-Williams maze; (2) free environment with other animals but devoid of playthings; (3) animals placed in a wire mesh cage, 8 in. wide X 6 in. high, within a free environment containing both animals and playthings; (4) mesh cages in free environment with peers but without playthings; (5) mesh cages in a free environment containing neither other animals nor playthings; (6) mesh cages in a small area environment with only playthings visually available; and (7) restricted environment, living individually in laboratory cages having three solid metal walls. Following 60 treatment days the animals were all housed individually, placed on a food deprivation regimen, and tested in the Hebb-Williams apparatus. The free environment plus plaything environment produced the best problem-solvers. The next best performance following rearing in the free environment was with other animals but without playthings. The mesh-reared groups did not differ from one another but were all superior to the restricted group. The results were interpreted as supporting the concept of the essential nature of early perceptual learning. Again, response transfer from the rearing experiences and possible adaptation bias confound this interpretation.

Bingham and Griffiths (1952) reared 48 rats under three conditions: (1) a free environment (6 ft X 9 ft) room where subjects received food in three different situations over a period of 30 days, i.e., by means of traversing a straight alley by running an inclined plane to a platform for food and by pushing through a swinging door; (2) individual housing in squeeze boxes (2 in. X 5 in. X 4 in.) constructed of hardware cloth; and (3) individual rearing in standard lab cages. After 51 days of rearing in their respective experimental conditions, the animals from groups 1 and 2 were housed in standard laboratory cages. All subjects were then adapted to a Warner-Warden maze, a linear arrangement of 12 two-alternative choice points, for a period of 3 days. Following adaptation, five trials were given daily until a criterion of 10 successive, errorless responses was recorded for each animal. Reinforcement consisted of access to a dish of wet mash for 15 sec. Following this testing, subjects were adapted to an inclined plane maze, which consisted of five levels, which required traversion from lower to upper levels in order to reach the goal box. An error was scored each time a subject failed to traverse the appropriate ramp to reach the goal. Each rat was run for 5 trials per day to a total of 40 trials. Finally, subjects were familiarized with the Lashley jumping stand for a period of 3 days, jumping through each window 10 times from distances of 11, 13, and 17 cm. The stimulus cards were rotated randomly to avoid position habits. Errors and trials to a criterion of 20 successive correct were recorded during the subsequent testing.

Squeeze-box and cage-reared animals were not found to differ statistically in performance and were thus combined to form a single control group. Their performance was compared to those of the animals reared in the "wider and richer environment of the experiment rooms." It is quite obvious that the feeding situations to which the experimental animals were exposed may have directly transferred responses to the testing situations, straight alley to Warner-Warden maze, ramp running to the inclined plane maze, and pushing through swinging doors to terminal card-pushing response in the Lashley jumping stand. In an attempt to partition these transfer effects a supplementary experiment was run. Ten rats were reared in a room containing alleys, inclined planes, and swinging doors, and they were subjected to feeding situations like those of group 1 noted above. An equal number of rats were reared in a similar room containing a broken chair, a box, and a broken cage. At 78 days of age these animals were tested in the Warner-Warden maze as described above.

In both experiments, subjects reared in the open rooms were superior to the restricted controls on the Warner-Warden maze with respect to both total errors and trials to criterion. On the inclined plane maze, used only in the first experiment, rats reared in the "richer" environment had fewer error scores (statistically significant) and more favorable time scores (not significant). In performance on the Lashley jumping stand, all encompassingly referred to as the "discrimination test," no significant differences were found. Elsewhere in the report we are told that the groups did not differ with respect to temperament or susceptibility to sound-induced seizures. It was concluded that richer environments facilitated "maze-learning ability." The notion of "wideness" characteristic of the enriched environment was seen as having a greater influence on adult behavior than the specific environmental encounters experienced in the feeding situations. The suggestion that the specific experiences rendered by the feeding situations did not affect performance during testing is unwarranted and may be deceptive. In the supplementary experiment described, performance was measured only on one maze, the Warner-Warden, and not "on the mazes" as was incorrectly summarized. Because the inclined plane maze and the Lashley jumping stand were not employed, the inclined planes and swinging doors had no transferable significance for the test situation.

Two prominent methodological difficulties are inherent in the above perceptual learning experiments: (1) perceptual deprivation is confounded with social isolation; and (2) subsequent learning tests are not directed toward the elucidation of any overall learning capacity as is suggested by their interpretation. Rather, the tests are so designed as to facilitate the transfer of specific rearing experiences engendered by particular rearing apparatus to postrearing learning tests.

The theoretical effects of relatively limited differences in early experiences upon subsequent learning were measured by Forgus (1954, 1955). Forgus compared the maze performance of rats reared in an open area containing three-dimensional objects around the margin, with a group reared in a similar environment except that a Plexiglas enclosure surrounded the margin preventing the animals from interacting physically with the objects. Forgus assumed that the open rearing condition stimulated both visual and physical interaction, while the closed condition primarily involved visual stimulation. It was predicted that if both groups were given maze training in an 11-unit T maze in a well-lighted room (both physical and visual cues available) the visual-reared rats would be superior since they would be less distracted by the physical cues. However, if the visual cues were markedly diminished (lights turned out), the visual-and-physical-reared rats would emerge superior. The results supported these predictions, with the visual-reared rats requiring fewer trials to reach initial criterion and with their behavior significantly more disrupted when the lights were turned out. However, some of the Forgus explanatory assumptions may be incorrect. First, the visual group was reared in a consistently smaller, and thus denser, area. Second, it was assumed that the two groups would be equal with respect to visual learning during rearing. This depends on the assumption that whether or not the objects could be contacted, they would elicit equal amounts of attention. Also it was assumed that the objects could be seen equally well by both groups. Thus, rather than relating the obtained group differences to general facilitation of learning ability by the rearing condition, it again seems more reasonable to attribute the effects to specific responses learned in the physical and visual condition. Because the test maze was elevated and since the rats reared in the physical-visual condition had had extensive experience climbing on the elevated objects in their rearing environment, perhaps they adapted more readily to the elevated nature of the testing condition.

Gibson and Walk (1956) investigated the dependence of visual form discrimination in adult rats on a specific variation in early visual stimulation. Experimental subjects were reared in groups of four from birth to 90 days with a mother in cages which had three-dimensional circles and triangles on the walls. These stimuli were identical to those used in later discrimination learning tests. The control group was raised under the same conditions but without opportunity to see the forms before discrimination learning. At 90 days the rats were placed on a 24-hr feeding cycle for 1 week. They were then tested for 15 days in a Grice box using circles and triangles as discriminative stimuli. The group exposed to the stimuli during development made fewer errors and required fewer trials to reach criterion. In this experiment, experimental subjects may have simply been identifying already familiar forms or a particular pattern, rather than exhibiting a general facilitative effect of their early experiences. Gibson, Walk, Pick, and Tighe (1958) attempted to clarify whether the observed effects were general or specific in nature. In one experiment rats were reared in (1) an experimental condition with three-dimensional equilateral triangle and circle placques on the cage walls, and discrimination learning of the same shapes beginning at 90 days; (2) a control condition in which the rats were reared without patterns on the cage walls and were tested on the triangle-circle discrimination at 90 days; (3) an experimental condition with triangles and circles on the walls from birth and training at 90 days on a discrimination between an elipse and an isosceles triangle; and (4) a control condition with no forms during rearing, followed by ellipse-isosceles triangle training at 90 days of age. Both experimental groups performed with fewer errors in discrimination training than their respective controls. The results suggested that what the animal learns from viewing the triangle and circle on the cage walls is not specific identification of these two patterns. However, this interpretation is questionable in that ellipses and isosceles triangles are simple transformations of circles and equilateral triangles. To further clarify the issue, Gibson *et al.* (1958) raised six groups of rats. The main experimental groups had triangles and circles on the cage walls. Group E_1 was tested on a triangle-circle discrimination at 90 days, while group E_2 learned a horizontal-vertical stripe discrimination. Two control groups had no wall patterns during rearing. Group C_1 learned the circle-triangle discrimination at 90 days, while group C_2 learned the horizontal-vertical stripe discrimination. Groups RE_1 and RE_2 were reared with irregularly shaped rocks hung on the walls. One group learned the triangle-circle discrimination at 90 days of age, while the other discriminated the striped patterns. As expected, the group reared with the circles and triangles showed significant facilitation during discrimination training using these patterns as compared to their control condition. No differences in rate of acquisition of number of errors occurred between any of the groups in the stripe pattern discrimination problems regardless of exposure during development. However, those rats who were reared with rocks on the cage walls performed as well as those rats reared with circles and triangles when tested on the circle-triangle discrimination problem. This provided some indication that rearing conditions may produce a general facilitatory effect on learning as opposed to transfer of specific response elements. Forgus (1958a) tested the hypothesis that positive effects from early perceptual exposure are based on familiarity and upon the extent to which early experiences with stimulus forms produce selective responding to the perceptual differences between the stimuli of the problem task. In other words, if the testing forms were similar to the exposure forms but contained distinct elements of novelty, animals would respond selectively to those novel elements. Hooded rats were placed at weaning in four groups. All experimental groups had identical three-dimensional circles at opposite ends of the rearing cage. Three-dimensional triangular forms were on the other two opposing walls. The triangles were different for the three experimental groups. Group T had a total triangle, group S had the sides but not the angles of the triangle, while group A was exposed to the angles of the triangle devoid of the sides. A control group was reared without visual access to any of these forms. It was assumed that the informational aspects of the triangle were concentrated at the angles. Thus, it was expected that group S would show the greatest facilitation in discrimination learning since the testing forms would be similar yet novel, thus directing preferential responding to the triangle. This expectation was confirmed. Forgus (1958b) also found that rats reared with the form of the total triangle responded with fewer errors when required to discriminate between a triangle without angles and a circle, compared to rats who were reared with the partial triangle.

Hymovitch (1952) suggested that there may be age boundaries after which exposure to an enriched environment would have no facilitative effect. He supposed that the treatment must occur before maturity. Forgays and Read (1962) investigated age parameters, rearing six groups of rats. The first group was exposed to an enriched environment (group housing, including ramps, barriers, and geometric objects) from birth to 21 days of age. Group 2 was exposed to the enriched environment from day 22 to 43, group 3 from day 44 to 65, group 4 from day 66 to 87, and group 5 from day 88 to 109. A sixth group was reared in a standard laboratory cage devoid of extra stimulus objects. All subjects were tested at 114 days of age in an elevated T maze measuring motor activity, and then run on 12 detour tests in the Hebb-Williams apparatus. No differences appeared among the groups in maze activity. Mean error scores on the Hebb-Williams tests revealed that rats exposed to the enriched environment at 22-43 days made significantly fewer errors than the nonenriched group. All animals receiving the enriched treatment were superior to the nonenriched group, with the magnitude of the effect greatest when the treatment was given soon after eye opening. More recently Nyman (1967) has found that exposure to any early enriched environment has its greatest facilitory effect on learning when the experience was provided between the ages of 50 and 60 days. These effects occurred on an alternation maze, the Hebb-Williams maze, but not on a T-maze visual discrimination. It should be noted that since these studies tested all groups at the same age, the age of early experience was confounded with time between end of treatment and start of testing.

Diffuse supernormal or enriched stimulation associated with handling also has been shown to enhance later learning. For example, Levine and Welzel (1963) handled half of each group of three rat strains (Harlan Long Evans, Rockland Long Evans, Sprague-Dawley) from birth to 90 days. Control rats were not handled. At 90 days each animal was tested in a shuttle-box avoidance-learning situation, a situation utilizing shock as an unconditioned stimulus. On the average the handled rats learned faster than the unhandled control animals. Among the various strains, the handled rats of the Long Evans strains learned faster than their control groups, but the two groups of Sprague-Dawley rats did not differ. It appears, then, that the effects of early experience may interact with genetic factors, so that conclusions concerning environmental effects require specification of the animal species and the specific test situation (Denny & Ratner, 1970).

Gill, Reid, and Porter (1966) reared rats singly in bare wire cages from day 21 to 81. It was found that these restricted rats performed as well as rats reared in an enriched environment when tested on visual discriminations and reversals presented in the Lashley jumping stand. Bennett and Rosenzweig (1969) reared rats in restricted, standard, and enriched environmental conditions. After 30 days of differential experience, subjects were tested on a successive visual reversal discrimination task. Isolated rats were inferior in problem-solving scores, but only if the restriction was begun immediately after weaning. No effects of the restriction were disclosed if the treatment was begun at 60 days of age.

Most of the modifications of learning ability reported in these studies have been associated with maze performance, predominantly the Hebb-Williams maze. It is interesting to note that performance of cats and rhesus monkeys in this maze does not improve with chronological age in either species, nor is the performance of monkeys superior to that of cats (Warren, 1965). Wilson, Warren, and Abbot (1965) suggested that performance on the Hebb-Williams maze may benefit specifically from past experience in open fields, a characteristic provided by most enriched environment rearing conditions. Woods et al. (1961) proposed that deficits in learning after deprivation rearing in subprimate mammals may actually involve motivational and perceptual, rather than intellectual, processes. They found that rats reared in isolation conditions were consistently more active when introduced into an open field. On most maze tests, such exploratory behavior and hyperactivity will yield performance errors by isolated rats. With regard to this interpretation, other investigators have found that either enriched rearing or restricted rearing may potentiate exploration depending on the specific testing situation (e.g., Forgus, 1954; Gill et al. 1966).

EFFECTS OF EARLY EXPERIENCE UPON LEARNING IN DOGS

Thompson and Heron (1954) compared the learning performance of 26 Scotch terriers reared under social isolation conditions or in an enriched environment. The enriched dogs were reared either as laboratory pets or as pets living in the homes of laboratory personnel.

The animals were tested on a battery of learning tests with apparent graded difficulty. One test involved running down a wall from a start corner to a food or goal corner. A partial reversal of this response was then established with the animal required to run to a goal corner diagonal to the first. A second test required the dogs to run from the middle of the open area to a goal corner followed by a reversal of the correct corner. The third and fourth tests were Hebb-Williams-like detour problems, first with a parallel barrier, then with an L-shaped barrier. The isolate-reared dogs made significantly more errors on the discrimination and discrimination reversal problems, and required more time to successfully solve the detour problems.

The final learning evaluation test was a delayed-response problem. In this problem the animals were shown the location of a food incentive by placing the bait under one of two identical containers differing only with respect to their position on the testing apparatus. Each subject was permitted to respond following the elapse of a certain length of time, including zero delays. The performance of Scotch terriers reared as pets in the homes of laboratory personnel was clearly superior to that of subjects reared in isolation.

We do not believe that any of the reported differences can be interpreted as loss in learning capacity. The experiences given the enriched dogs could very well have transferred directly to the test situations. For example, the experience of the enriched dogs to various "naturally" occurring open fields should have provided preadaptation to the tests involving modified open fields. With respect to the delayed-response tasks, the pet rearing of dogs most likely includes situations where the dogs are trained to wait upon the master's command. Furthermore, pet-raised dogs should be under less emotional pressure than isolate-reared dogs in the delayed-response situations, and delayed-response performance is greatly influenced by stress.

Melzack and Scott (1957) reared 10 terriers in total social isolation and 12 dogs as pets in both private homes and the laboratory. Two of the reported tests did not involve learning as such, but are pertinent to this discussion. The animals' response to burning and pin pricking was tested. Seven out of 10 isolated dogs made no attempt to avoid a flaming match pushed into their noses, while all of the pet-reared dogs demonstrated rapid and effective flame avoidance. Another test involved stabbing the dogs in their flanks with dissecting needles. Pet-reared dogs spent little time with the experimenter following this treatment, but most isolate dogs increased time spent proximal to the experimenter following the pin pricking.

The experimenters tested the two groups on shock-avoidance training. It was found that the free-environment dogs acquired the avoidance response significantly faster than the isolated dogs. The protocols of the isolates' test performance indicate that they consistently froze in response to the presentation of the shock. Certainly this response successfully competed with the appropriate instrumental response and therefore precludes any interpretation of these data involving the notion of learning capacity.

Fuller (1966) reported that beagles reared in isolation were generally poor performers when tested on a series of spatial reversals presented in a modified Y maze. However, between-subject variability was very large, with some isolates outperforming the controls. Fuller (1967) found that isolated dogs did as well as pet-reared dogs in the development of a visual discrimination. However, when a series of reversals was instituted isolate performance deteriorated, but not significantly more than the pet-reared dogs.

Lessac and Solomon (1969) studied the effects of deprivation rearing on learning in beagles. A research design was introduced which required both pre- and postisolation testing. The experimenters expected this design to enable them to differentiate between developmental failure and deterioration interpretations of isolation effects. The evaluative tests included Hebb-Williams detour problems, avoidance training, and classical conditioning. It was reported that judging from the performance of the dogs before isolation it appeared that capabilities were lost during the isolation period. The investigators stated, "behavioral deficiencies found in isolated subjects will not reflect a simple slowing of normal development, but must represent an active destructive atrophic process produced by the isolation experience." To discover that pre- and postisolation learning performances are different, does not necessarily validate the notion of *deteriorated capacities*. Unless it can be demonstrated that the levels of behaviors that might successfully compete with learning performance have not changed from pre- to postisolation periods, the notion of capability changes is devoid of any practical significance. In fact, Lessac and Solomon described in great detail the enormous emotional disturbance of their beagles when they emerged from isolation. Further, normative maturational data necessary for adequate interpretation was not provided. It is safe to say that their interpretations are more than suspect.

EFFECTS OF EARLY EXPERIENCE UPON LEARNING IN MONKEYS AND APES

The extensive research carried out by Harlow and his associates at the Wisconsin primate laboratories relevant to this discussion will be reviewed in another chapter. Suffice it to say that, in general, no significant differences had been found between isolates and controls with respect to the Wisconsin General Test Apparatus (WGTA) test battery at the time the chapter was written.

Angermeier, Phelps, and Reynolds (1967) studied four groups of rhesus monkeys subjected to either total isolation, partial isolation, living in pairs in a small cage, or pair-living in a large cage containing play objects, colored panels, and rotating lights affixed to the wall. Following these differential experiences the animals were tested for discrimination learning on a four-choice match-to-sample series. The forms were projected through a clear space in the performance panel of an operant chamber. Subjects were either positively reinforced with food for correctly matching the projected sample or shocked if an incorrect match was made. No effects of the differential experiences were observed over the extended period of testing employed (6000 trials). The generality of these particular negative results must be questioned on a number of grounds. First, unlike most primate studies where the subjects are exposed to the particular rearing conditions at birth, the subjects in this experiment were obtained from an importer and placed in their rearing conditions at 2 months of age. Secondly, the animals were kept in their rearing conditions for only 2 months. It had been demonstrated previously (Griffin & Harlow, 1966) that 3 months of total isolation is not sufficient to produce lasting effects, even with respect to highly vulnerable social behaviors. Thus the described period of social deprivation would not be expected to produce learning deficits.

Gluck (1970) trained isolates, partial isolates, and peer-reared monkeys to operate a manipulandum for sucrose reinforcement. Adaptation and acquisition criteria found that isolates required more time to take food from a magazine, took more time to emit the initial lever press, but did not differ from the other groups in acquisition of the lever response once the initial lever press was made. It was found that isolates tended to respond faster than the other animals when they were placed on a continuous reinforcement schedule. Following this phase, subjects were tested on an alternating extinction and reacquisition schedule. The isolates emitted significantly more responses during the first several extinction components. Though the results are statistically reliable, it would be premature to attribute these apparent perseverative tendencies to some notion of altered intellectual capacity.

It has been our intention in this paper to question the uncritical acceptance of the concept that deprivation rearing alters basic learning abilities. There can be no doubt that the primary effect of early deprivation is the alteration of emotional and temperamental variables, and if these cannot be ruled out, the effects of early deprivation cannot be determined.

It is absolutely essential that postrearing learning tests are not motivated by painful stimuli, since it is a distinct possibility that deprived and enriched animals differ in terms of their responsiveness to these stimuli.

We strongly believe that it is of limited scientific value to give enriched animals extensive experience in open fields during rearing and then compare their performance in modified open fields with animals deprived of such experience.

An almost perfect way to confound all early experience variables is that of raising the enriched subjects given free rein of a laboratory or raised in human home. The human home is an idealized open field, it guarantees maximal contact and emotional desensitization of human experimenters and it offers opportunity for an unlimited number of unspecified specific learning experiences. These facts leave many of the rodent and dog experiments relatively uninterpretable.

Providing the enriched subjects with specific components of the apparatuses or stimulus displays to be used later to assay differential effects of deprived and enriched environments on learning is not uncommon and is an entirely inadequate experimental procedure.

Because of the differential emotional status of subjects emerging from enriched and deprived environments all subjects must exhibit equal adaptation to all learning situations. Adequate adaptation of subjects raised in deprived environments to a single-test situation is difficult and adaptation to multiple-test situations is probably impossible. Most experimenters either totally ignored or gave short shrift to the problem of test situation adaptation.

It is entirely possible that the discrepancies in the literature on the effects of deprived and enriched early environments on later learning ability stem from the fact that the conception of the problem has changed with time. The early literature suggests a desire to demonstrate that there are differences in learning performance in animals raised in enriched or deprived environments whether or not these differences were related to differential perceptual experience obtained from the early environments, whether or not these learning differences were the result of emotional or temperamental differences produced by the early environments or to demonstrate that limited and specific learning experiences given early in life would generalize and produce significant differences in learning performance in adulthood.

Strictly speaking there is nothing improper about labeling these kinds of effects as intellectual changes as long as the experimenter clearly reports the contribution of all known salient variables. It should be noted that the concept of intelligence or ability is not without its surplus meaning, and it is thus essential for the experimenter to make perfectly clear the kinds of behaviors he personally subsumes under these headings.

REFERENCES

Angermeier, W. F., Phelps, J. B., & Reynolds, H. H. The effects of differential early rearing upon discrimination learning in monkeys. *Psychonomic Science*, 1967, **8**, 379-380.

Bennett, E. L., & Rosenzweig, M. R. Potentials of an intellectually enriched environment. In (Chm.), Dysnutrition in the seven ages of man. Symposium presented at the meeting of San Francisco, 1969.

Bingham, W. E., & Griffiths, W. J. The effect of differential environments during infancy on adult behavior in the rat. *Journal of Comparative and Physiological Psychology*, 1952, **45**, 307-312.

Denny, M. R., & Ratner, S. C. Behavioral consequences of early behavior: Early experience. In *Comparative psychology: Research in animal behavior.* Homewood, Illinois: Dorsey Press, 1970.

Forgays, D. G., & Forgays, J. W. The nature of the effect of free-environmental experience in the rat. *Journal of Comparative and Physiological Psychology*, 1952, **45**, 322-328.

Forgays, D. G., & Read, J. M. Crucial periods for free-environmental experience in the rat. *Journal of Comparative and Physiological Psychology*, 1962, **55**, 816-818.

Forgus, R. H. The effect of early perceptual learning on the behavioral organization of adult rats. *Journal of Comparative and Physiological Psychology*, 1954, **47**, 331-336.

Forgus, R. H. Early visual and motor experience as determiners of complex maze-learning ability under rich and reduced stimulation. *Journal of Comparative and Physiological Psychology*, 1955, **48**, 215-220.

Forgus, R. H. The effect of different kinds of form pre-exposure on form discrimination learning. *Journal of Comparative and Physiological Psychology*, 1958, **51**, 175-178. (a)

Forgus, R. H. The interaction between form pre-exposure and test requirements in determining form discrimination. *Journal of Comparative and Physiological Psychology*, 1958, **51**, 588-591. (b)

Fuller, J. L. The K-puppies. *Discovery*, 1964, **25**, 18.

Fuller, J. L. Transitory effects of experimental deprivation upon reversal learning in dogs. *Psychonomic Science*, 1966, **4**, 273-274.

Fuller, J. L. Experiential deprivation and later behavior. *Science*, 1967, **158**, 1645-1652.

Gerall, H. D., Ward, I. L., & Gerall, A. A. Disruption of the male rat's sexual behavior induced by social isolation. *Animal Behaviour*, 1967, **15**, 54-58.

Gibson, E. J., & Walk, R. D. The effect of prolonged exposure to visually presented patterns on learning to discriminate them. *Journal of Comparative and Physiological Psychology*, 1956, **49**, 239-242.

Gibson, E. J., Walk, R. D., Pick, H. L., & Tighe, T. J. The effect of prolonged exposure to visual patterns on learning to discriminate similar and different patterns. *Journal of Comparative and Physiological Psychology*, 1958, **51**, 584-587.

Gill, J. H., Reid, L. D., & Porter, P. B. Effect of restricted rearings on Lashley Stand performance. *Psychological Reports*, 1966, **19**, 239-242.

Gluck, J. P. Successive acquisitions and extinctions of bar-pressing: The effects of differential rearing in rhesus monkeys. Unpublished master's thesis, University of Wisconsin, 1970.

Griffin, G. A., & Harlow, H. F. Effects of three months of total social deprivation on social adjustment and learning in rhesus monkeys. *Child Development*, 1966, **37**, 534-547.

Harlow, H. F., & Harlow, M. K. The affectional systems. In A. M. Schrier, H. F. Harlow, & F. Stollnitz (Eds.), *Behavior of nonhuman primates*. Vol. II. New York: Academic Press, 1965.

Hebb, D. O. *The organization of behavior*. New York: Wiley, 1949.

Hebb, D. O., & Williams, K. A method of rating animal intelligence. *Journal of General Psychology*, 1946, **34**, 59-65.

Hymovitch, B. The effects of experimental variations on problem solving in the rat. *Journal of Comparative and Physiological Psychology*, 1952, **45**, 313-321.

Lessac, M. S., & Solomon, R. L. Effects of early isolation on the later adaptive behavior of beagles: A methodological demonstration. *Developmental Psychology*, 1969, **1**, 14-25.

Levine, S., & Welzel, A. Infantile experiences, strain differences and avoidance learning. *Journal of Comparative and Physiological Psychology*, 1963, **56**, 879-881.

Melzack, R., & Burns, S. K. Neurophysiological effects of early experience. *Experimental Neurology*, 1965, **13**, 163-175.

Melzack, R., & Scott, T. H. The effects of early experience on the response to pain. *Journal of Comparative and Physiological Psychology*, 1957, **50**, 155-161.

Miller, R. E., Caul, W. F., & Mirsky, V. Communication of affects between feral and socially isolated monkeys. *Journal of Personality and Social Psychology*, 1967, **7**, 231-239.

Nyman, A. J. Problem solving in rats as a function of experience at different ages. *Journal of Genetic Psychology*, 1967, **110**, 31-39.

Riesen, A. H. Sensory deprivation. In E. Stellar, & J. M. Sprague (Eds.), *Progress in physiological psychology*. New York: Academic Press, 1966.

Rosenzweig, M. R. Environmental complexity, cerebral change, and behavior. *American Psychologist*, 1966, **21**, 321-332.

Sackett, G. P. Effects of rearing conditions upon the behavior of rhesus monkeys. *Child Development*, 1965, **36**, 855-868.

Sackett, G. P. Innate mechanisms, rearing conditions, and a theory of early experience effects in primates. In M. R. Jones (Ed.), Miami Symposium on the prediction of behavior: Early experience. Coral Gables: University of Miami Press, 1970.

Scott, J. D. Critical periods in behavioral development. *Science*, 1962, **138**, 949-958.

Thompson, W. R., & Heron, W. The effects of restricting early experience on the problem solving capacity of dogs. *Canadian Journal of Psychology*, 1954, **8**, 17-31.

Warren, J. M. The comparative psychology of learning. *Annual Review of Psychology*, 1965, **16**, 95-118.

Wilson, M., Warren, J. M., & Abbott, L. Infantile stimulation, activity, and learning by cats. *Child Development*, 1965, **36**, 843-853.

Woods, P. J., Fiske, A. S., & Ruckelshaus, S. I. The effects of drives conflicting with exploration on the problem solving behavior of rats reared in free and restricted environments. *Journal of Comparative and Physiological Psychology*, 1961, **54**, 167-169.

CHAPTER 6

The Effect of Early Adverse
and Enriched Environments
on the Learning Ability
of Rhesus Monkeys[1]

H. F. Harlow, M. K. Harlow, K. A. Schiltz,
and D. J. Mohr

HISTORY

Since deepest antiquity man has speculated about the effects of various kinds of early environments, ranging from those completely restricted to those enormously enriched, on the subsequent personal-social and intellectual developmental capabilities of human beings and other animals. Presumably, early restricted childhood environment impairs later intellectual development, and early enriched environment enhances later intellectual development.

One of the time-honored tales involving environmental restriction is that of Romulus and Remus, who as infants were reportedly raised by a wolf. Even a well-meaning maternal wolf would provide a restricted environment for a human being. Since Romulus subsequently became a successful Roman emperor he apparently suffered no irreparable social or intellectual damage. Actually, it is physically and physiologically impossible for a wolf to rear a human, but people have "cried wolf" before, and people will "cry wolf" again. This is equally true for those conjectured, cuddling, canine creatures whether they live in Africa (Foley, 1940a, 1940b) or India (Squires, 1927; Singh & Zingg, 1942; Bettelheim, 1959; Ogburn, 1959; Ogburn & Bose, 1959).

There is historical evidence that a Bavarian prince, Kasper Hauser (Wasserman, 1928), was placed in a single-room dungeon as a very young child and saw no other person and nothing of the wondrous, wide world beyond until, or

[1] This research was supported by USPHS grants MH-11894 and RR-0167 from the National Institutes of Health to the University of Wisconsin Primate Laboratory and Regional Primate Research Center, respectively.

121

shortly before, puberty. In the few years after he was released from captivity and before he was murdered, Kasper Hauser was reported to have made an admirable social and intellectual recovery. Alas, his Stanford-Binet, Wechsler, Rorschach, TAT, and Minnesota Multiphasic Personality Inventory scores were never recorded, and Kasper Hauser will remain an enchanting legend rather than a well-documented clinical case.

Fortunately, Itard (1932) has left us a magnificent record of creative testing on the Wild Boy of Aveyron, whom Itard studied over a number of years. This boy, the "idiot of Itard," never recovered normal personal-social or intellectual abilities subsequent to his early experience, but there is every reason to believe that he had not been very bright initially. In all probability he was born with less than an average intellect, and it is unlikely that there ever was a wolf. Perhaps the French wolves were so busy eating children during the 19th century that they had no time to nurture them. The only scientific contribution made by study of the wild boy was the tests Itard created, which included a form of the delayed response.

Interest in the effects of early deprived environments on human beings was kindled by the researches of Spitz, who reported that children raised in restricted early environments, as typified by subjects from two very inadequate orphanages, developed marasmus (Spitz, 1946) and hospitalism (Spitz, 1945), syndromes which were dramatized by disturbed and distorted personality development and significant developmental loss. Actually Spitz's developmental quotient, derived from the Hetzer-Wolf (not a real wolf) baby tests, did not adequately measure learning ability, and the Hetzer-Wolf is neither a reliable nor a valid predictor of adult intelligence. So far as the reported dulling, and even destruction, of personal-social traits is concerned, we believe that Spitz was basically right.

During the last 10 years a series of long-term tests has been conducted at Wisconsin on the effects of various kinds of deprived, normal, and enriched environments on the personal-social and intellectual development of monkeys (Harlow, 1959; Harlow & Harlow, 1962; Harlow, Rowland, & Griffin, 1964; Rowland, 1964; Mitchell, Raymond, Ruppenthal, & Harlow, 1966; Harlow, Schiltz, & Harlow, 1969). We have differentiated between sensory and social deprivation and between total and partial social deprivation. In our researches we have tried to avoid sensory deprivation since early sensory deprivation, especially total visual deprivation, is known to produce degenerative changes in the central nervous system.

Total social isolation is defined as complete absence of any social contact, particularly social contact with members of the animal's species, from birth until a predetermined time (Harlow et al., 1964; Rowland, 1964; Griffin & Harlow, 1966). In partial social isolation, animals are permitted to see and hear members of their species, but are denied the opportunity to physically interact with them (Harlow & Harlow, 1962). There are, of course, an unlimited number of special partial social deprivation conditions, such as maternal rearing with peer deprivation (Alexander, 1966), peer rearing with maternal deprivation (Chamove, 1966; Harlow, 1969), surrogate mothering combined with relatively normal peer experience (Hansen, 1962), and even father deprivation, a condition almost universal in all primate laboratories before the creation of the nuclear family situation.[2]

By and large, early social deprivation exerts dire and devastating effects upon the personal-social abilities of macaque monkeys, and within definable limits there are critical periods for these effects. Since these "critical periods" take place over weeks and months, not days or hours, the term is meaningless if defined in the Lorenz "imprinting manner" (Lorenz, 1935), where learning is instantaneous, totally specific, and ever-unchanging. The effects on primates are gradual and even reversible for a considerable period of time, thus a better term is "sensitive periods."

Monkeys subjected to total social isolation during the first 3 months of life make a rapid and complete social recovery when subsequently allowed to interact with age-mates (Griffin & Harlow, 1966). Monkeys totally socially isolated for the first 6 months of life subsequently make inadequate social adjustments (Harlow & Harlow, 1962; Rowland, 1964), and since their abnormalities persist in very large part for more than 3 or 4 years after they have been given opportunities to socialize (Mitchell et al., 1966), one may properly presume that permanent loss of socialization capabilities has resulted.

Thus, the sensitive period of socialization for macaque monkeys lies between the first 3 to 6 months, or perhaps 3 to 12 months, of life, recognizing that this is an "elastic" period that cannot be specified in hours, days, or weeks. Possibly the sensitive period cannot be specified adequately even in terms of months, since 1 year of total social deprivation produces even more social devastation than does a half-year, and since successful or partially successful attempts at rehabilitation have been initiated only recently.

The devasting effects of protracted total social isolation in primates cut across all facets of social behavior, and the severity probably scales with deprivation duration. Socially deprived monkeys make no attempt to even seek social contact, as shown in Fig. 1, and if they are relatively young, social isolates make no attempt at normal play. Socially deprived males may at maturity attempt sexual contacts, but these efforts, as shown in Fig. 2, are futile, funny, and frustrating. The female is commonly sexually incompetent and indifferent, as shown in Fig. 3, even though basic sexual reflexes such as sexual-present, depicted in Fig. 4, persist.

Socially deprived monkeys, as illustrated in Fig. 5, are terrified by both age-mates and younger monkeys. Fear normally masks aggression, but social isolates when they mature are hyperaggressive, or transiently hyperaggressive; 6-month isolates at 1 year of age may abuse even more hopeless and hapless 12-month isolates

---

[2] M. K. Harlow, unpublished data, 1970.

FIG. 1. Lack of social contact by socially isolated monkeys.

FIG. 2. Inadequate sexual postures by socially isolated male monkeys.

FIG. 3. Inadequate sexual posture by female monkey after social isolation.

FIG. 4. Sexual present by socially isolated female.

FIG. 5. Terror by older social-isolate monkeys in presence of normal infant.

and may attack a helpless infant or neonate—a behavior almost nonexistent in normal adolescent and adult monkeys (Mitchell *et al.*, 1966).

The subprimate literature possesses a wealth of massive, motivated, and often meaningless researches presumably bearing on the effects of early environmental deprivation on subsequent learning performance in rats and dogs (Krech, Rosenzweig, & Bennett, 1962; Thompson & Heron, 1954; Melzack & Scott, 1957; Lessac & Solomon, 1969). The primary contribution of these earlier researches is the illustration of errors in experimental design and procedure. However, these researches are described elsewhere, and this paper is limited to the role of various early environments on learning by macaque monkeys.

INTRODUCTION

The present research was designed to measure the effects of various types of early experience upon the subsequent learning ability of rhesus monkeys.

Groups of monkeys were subjected to total social isolation from a few days after birth until 6 or 9 months of age, and their learned performances were compared with those of control groups closely matched for age.

Additional groups of young macaques were raised in a socially enriched laboratory environment, called the nuclear family enrivonment, in which all infants had access to four adult females, one of which was its own mother, four adult males, four age-mate infants, and in some cases other younger infants. Again the learned performances of these macaques were compared with a large control group.

METHOD

Subjects

Forty-five laboratory-raised rhesus monkeys (*Macaca mulatta*) served as subjects and controls for this experiment. Twelve animals were denied all social experience during infancy (groups I-1, I-2, I-3), 12 were raised in socially enriched environments (groups E-1, E-2, E-3), and 21 were control subjects (groups C-1, C-2, C-3). The monkeys ranged in age from less than a year to over 2 years at the initiation of learning testing, but the age ranges of the experimental and selected control groups were comparable in that representative experimental and control groups were tested at each age level. Furthermore, no group was run on any test until attaining or closely approximating an age consonant with maximal adult learning efficiency (Harlow, 1959; Harlow, Blomquist, Thompson, Schiltz, & Harlow, 1968; Harlow, Thompson, Blomquist, & Schiltz, 1970). The complex learning test battery we used required approximately 18 months for completion, not including adaptation.

Two of the social isolate groups (I-1 and I-2, numbering five and four monkeys, respectively) were maintained under standard total social isolation conditions (Rowland, 1964) throughout the first 9 months of life, and a third group of three animals (I-3) was similarly maintained for a 6-month period.

The 12 monkeys that formed the three socially enriched groups were raised from birth in groups of four in our nuclear family housing facility. This situation is an enlarged and modified version of the standard playroom situation (Rosenblum, 1961; Harlow & Harlow, 1962).

The 21 control monkeys were placed in groups of 5, 4, and 12 animals housed individually in standard laboratory cages 28 in. X 28 in. X 28 in. where they could see and hear other laboratory animals but could not physically interact with them, as shown in Fig. 6. We have described this situation as partial social isolation (Harlow & Harlow, 1962), but monkeys so raised perform as well on standard learning tests as monkeys given ample opportunity for varied types

FIG. 6. Monkeys in partial social isolation.

of social interaction. Furthermore, Singh (1969) has reported data showing that both forest-raised and urban-raised monkeys in India perform in an almost identical manner to our control monkeys when tested on the standard learning tests subsequently described.

Apparatus

The monkeys subjected to total social isolation were raised from a few hours after birth until 6 or 9 months of age in social isolation cages designed by

Rowland (1964) and Sackett (1966). The Rowland apparatus, illustrated in Fig. 7, was a chamber 24 in. X 24 in. X 24 in. Three sides of the cubicle were sheet metal and the floor was wire mesh. Steel bars were placed 2 in. apart across the fourth side of the cubicle. An opaque fiberboard screen partitioned the living chamber from the enclosed testing area, which was not used in the present experiment. By raising the forward opaque screen the experimenter could observe the monkeys through a viewing port and could record their behaviors. Figure 8 shows a 6-month isolate monkey immediately after the forward screen was first raised. The total isolation apparatus used by Sackett was a wire cage 24 in. X 24 in. X 28 in. Three walls and the ceiling were covered by Masonite and the area below the floor was an aluminum panel. The rear wall was a nonreflecting screen of ground glass on which visual stimuli could be projected.

The basic physical setup of the nuclear family situation consisted of a large central play area 8 ft X 7 ft X 4 ft, accessible at all times to the infants only, with four connected living cages 4 ft X 7 ft X 4 ft. Each living cage housed a mother with one infant and also a male consort or father. In four cases the living cage and play area also housed a younger sibling of the infant being tested. Because of the physical size of this test situation and the wealth of social interactions that continuously transpired from dawn past dusk, particularly

ISOLATION UNIT

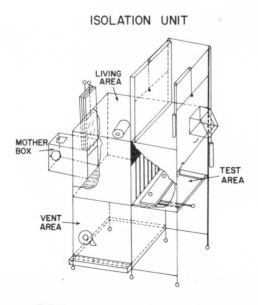

FIG. 7. Total social isolation apparatus.

FIG. 8. Monkey after release from 6 months of total social isolation.

when the overhead lights automatically dimmed, the experimental subjects enjoyed a great amount of social interaction and an opportunity to learn subtle social nuances seldom before provided laboratory-raised monkeys. Two pictures showing details of the apparatus and the infant inhabitants are presented in Figs. 9 and 10.

The learning test apparatus was the standard Wisconsin General Test Apparatus (WGTA) (Harlow, 1959), with appropriate trays and stimulus objects for the various problems. A portable WGTA was moved into a nuclear family environ-

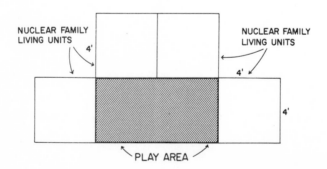

FIG. 9. Floor plan of nuclear family situation.

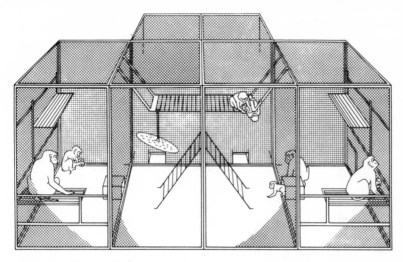

FIG. 10. Monkey interactions in nuclear family apparatus.

ment and used to adapt and test the monkeys in group E-3. Figure 11 shows a single trial setup for an object discrimination learning set problem and Fig. 12 shows an oddity learning-set problem.

FIG. 11. Response to discrimination problem by rhesus monkey.

FIG. 12. Response to oddity problem by rhesus monkey.

Procedure

Standard Adaptation Sequence. All subjects underwent a standardized adaptation sequence involving four basic stages, with the one exception that the enriched groups did not receive home-cage adaptation because they had been frequently hand fed by various people and thus adapted to hand feeding by the experimenter without strain or effort.

Home-Cage Adaptation. This adaptation phase involved home-cage feeding until the monkey accepted 25 pieces of food per session, usually within a 15-min period, for 5 consecutive days.

WGTA Adaptation. Using a single foodwell tray, we defined the adaptation criterion as acceptance of 25 pieces of food in a 10-min period, first with the well uncovered, later with the well partially covered by an unpainted wood cube using successive approximations, and then with the well completely covered. Finally, during a screen adaptation phase, the monkey was required to displace the object and accept food on 25 trials within 10 min, for each of 5 successive days. The forward opaque screen was closed when the trial began, and the back, one-way vision screen was down when the trial was conducted.

Object Position Adaptation in the WGTA. This phase involved displacement of a single object on a two-foodwell form board 25 times in a 10-min period for 5 consecutive days, during which the object was shifted in a balanced, irregular positional sequence.

Single Object Series in the WGTA. This stage consisted of displacement 25 times in a 10-min period of a single baited object which was shifted from one foodwell to another in a balanced, irregular positional sequence.

Standard Test Battery. After completion of the standard adaptation series the socially deprived subjects and their controls underwent a battery of learning tests which included a series of 20 25-trial discrimination problems with one problem presented daily; 600 6-trial discrimination learning set problems with four problems presented daily; 90 days of 0- and 5-sec intermixed delayed-response problems presented at the rate of 20 trials daily; 60 days of intermixed 5-, 10-, 20-, and 40-sec delayed-response problems presented at the rate of 32 trials daily; and finally, 256 6-trial oddity problems presented at the rate of 8 problems daily.

Test Problem Sequence. Upon completion of adaptation the subjects and their controls were tested on the same problems, in the same order, with the exception that the delayed-response problems were run before learning set for the socially enriched monkeys and their controls, and the delayed-response problems were run after learning set for the socially deprived monkeys and their controls. The latter sequence has been the "normal order" in most previous studies.

There were two reasons for the inversion of the order of problem running. Some members of the enriched-environment groups and controls were under 1 year of age when postdiscrimination testing was begun (see Table 2). Our earlier maturation of learning data indicated that monkeys 200 days of age could solve delayed-response problems at a nearly adult level. Contrariwise, learning-set problems required about 360 maturational days for guaranteed efficient performance. Furthermore, we already had a thoroughly tested control group that ran the test battery in this sequence starting at about 290 days. These subjects had been a control group for multiple operated groups, studying age effects of various induced frontal lesions, and the problem sequence chosen was the only meaningful order for rapidly assessing delayed-response loss deficits, the primary loss inflicted by frontal lesions in the earlier studies.

Careful analyses of positive and negative transfer effects indicated that these effects were transient and trivial, if they existed at all. Actually, the only transfer effect found for sequential problem order was a slight negative transfer between discrimination learning set and oddity learning set when the discrimination learning set immediately preceded the oddity learning set. The sequential order in running the various animals of the three rearing conditions is given in

Table 1 and the present state of test completion of all groups is presented in Table 2.

TABLE 1. Problems and Sequential Running Order of Isolated and Enriched Monkeys

	Discrimination	Discrimination learning set	0-5 sec delayed response	Multiple delayed response	Oddity learning set
Isolated	1	2	3	4	5
Enriched	1	4	2	3	5

TABLE 2. Problems Completed by Isolated, Enriched, and Control Monkeys

	Discrimination	Discrimination learning set	0-5 sec delayed response	Multiple delayed response	Oddity learning set
Isolated-1_9	C	C	C	C	C
Isolated-2_9	C	½C			
Isolated-3_6	C	C	C	C	C
Control-1	C	C	C	C	C
Control-2	C	½C			
Enriched-1	C	C	C	C	½C
Enriched-2	C	½C	C	C	
Enriched-3	C		C	C	
Control-3	C	C	C	C	C

Test Procedure

Discrimination Problems. The monkeys were first tested on a series of 20 discrimination problems of 25 trials each. Twenty problems were chosen, since we have long known that a single discrimination has little or no reliability. We have found that if the correct stimulus is initially preferred the task may be solved without errors, but if the correct stimulus is initially nonpreferred, solution of a problem may not be achieved within a long training sequence (Harlow, 1959; Harlow, Harlow, Rueping, & Mason, 1960).

Analysis of the solution of 20 discrimination problems, each run for 25 trials in an experiment measuring the effects of bilateral frontal lesions in monkeys (Harlow et al., 1968), clearly revealed that interproblem learning was involved in that the last 10 problems were solved more rapidly than the first 10 problems. Thus, though we describe this test as discrimination learning it is in fact a simplified discrimination learning-set task.

Delayed-Response Problems. The second problem type given to the isolate monkeys, and given later in the problem sequence for the enriched environmental monkeys, was a battery of 0- and 5-sec delayed-response problems. Ten problems at each delay level were conducted for 90 days, a total of 900 0-sec and 900 5-sec problems. This long series of short delays was designed to adapt the monkeys to the disturbing delayed-response procedures and also to measure basic delayed-response capabilities.

After the initial delayed-response trials were completed the monkeys were tested on a composite delayed reaction series of 5-, 10-, 20-, and 40-sec delays with eight problems at each of the four delay intervals presented on each day until a total of 480 tests at each delay interval had been completed.

Discrimination Learning-Set Problems. A protracted series of 600 discrimination learning-set problems was given before any of the delayed-response tests for the deprived monkeys and their controls and after the composite delayed-reaction series for the enriched monkeys and their controls. Each problem was run for six trials, four problems per day for 150 test days, giving a total of 600 problems.

Oddity Learning-Set Problems. The final test given to all the subjects was a series of oddity learning-set problems. Each oddity problem was presented for six trials, using a noncorrection method. The problems were presented as simplified oddity tasks with the correct stimulus always being either at the extreme right or left position and never over the center foodwell. Eight oddity problems were run on each test day for 32 test days, a total of 256 problems.

Specific Test Procedures

The detailed specific trial procedures for the discrimination problems, learning-set problems, delayed-response problems, and oddity problems have already been described in detail (Harlow, 1959; Harlow et al., 1970).

Method of Data Analysis

Each of the following measures—adaptation, object discrimination, discrimination learning set, combined 0- and 5-sec delayed response, multiple delayed response, and oddity learning set—were subjected to statistical analyses. For each measure, except for the delayed-response and oddity tests, one-way analysis of variance with rearing condition (isolation-control-enrichment) as the independent variable was performed. Where the analysis revealed a significant group effect, the Fisher l.s.d. statistic was subsequently employed to determine specific group difference. As an independent analysis for each measure the Mann-Whitney U test was performed to obtain a second measure of paired group differences.

RESULTS

Adaptation

The average number of days required to complete adaptation by all subjects is given in Fig. 13 for total adaptation days and in Fig. 14 for WGTA adaptation

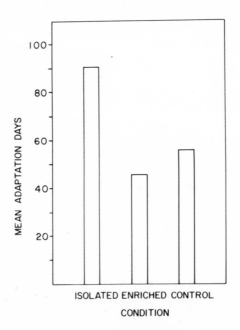

FIG. 13. Total adaptation days by isolated, enriched, and control monkeys.

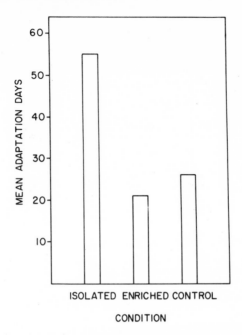

FIG. 14. WGTA adaptation days by isolated, enriched, and control monkeys.

days. The original category of home-cage adaptation was omitted in these analyses since the enriched-environment monkeys were already adapted to home-cage feeding before this experiment began. It is obvious that the isolated-condition animals took longer to adapt by both of the adaptation measures than did either the enriched-environment animals or the controls. However, the differences in adaptation time between the isolates and enriched monkeys resulted in large part from long adaptation time by a single isolate group and extremely short adaptation time by a single enriched group.

One-way analysis of variance with rearing condition (isolate-control-enriched) as the independent variable for the WGTA adaptation days yielded a significant condition effect (F = 7.51, df = 2/38, p < 0.01). The Fisher l.s.d. statistic was subsequently employed and revealed that the animals in the isolated condition were significantly inferior to both the enriched and control conditions at the 0.05 level. No significant differences between any of the three conditions were obtained by the Mann-Whitney test.

Object Discrimination

The object discrimination problems gave unexpected data in that the isolate monkeys were superior, closely followed by the control animals. Much to our

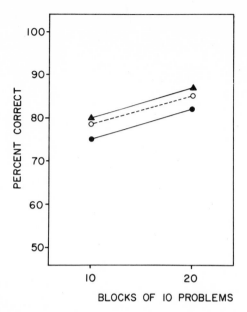

FIG. 15. Object discrimination performance by isolated, enriched, and control monkeys. (▲) 13 isolated, 676 days test age; (●) 12 enriched, 256 days test age; (○) 21 control, 484 days test age.

surprise the slowest learning on this test was shown by the enriched-environment monkeys. However, the three learning curves were closely parallel, as indicated in Fig. 15, and an unweighted means analysis of variance with rearing condition (isolate-control-enriched) as the independent variable failed to yield a significant condition effect ($F = 2.3$, $df = 2/43$, $p < 0.1$). However, a significant learning effect across trials was obtained.

Discrimination Learning Set

The learning set performances are plotted for the first 400 problems only, since group E-3 has not yet completed the final third of the problems. The data to date, as plotted in Fig. 16, revealed essentially identical abilities for monkeys in the control and isolate conditions, with both conditions superior to the enriched-environment monkeys.

Unweighted means analysis of variance with the three rearing conditions as the independent variable yielded a significant conditon effect ($F = 10.5$, $df = 2/26$, $p < 0.01$). The Fisher l.s.d. statistic subsequently showed the enriched condition to be inferior to both other conditions at the 0.05 level. Likewise, the

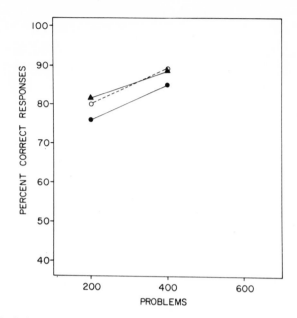

FIG. 16. Discrimination learning set by isolated, enriched, and control monkeys (trials 2-6). (▲) 12 isolated, 706 days test age; (●) 8 enriched, 501 days test age; (○) 21 control, 602 days test age.

Mann-Whitney test indicated the inferiority of the monkeys in the enriched condition to both other conditions at the 0.01 level. Again, a significant effect across blocks of trials was found.

Delayed Response, 0- and 5-Second

The performances of the three rearing conditions of monkeys on the short 0- and 5-sec delays are depicted in Fig. 17. The data for one isolate group, I-2, are incomplete and are omitted. The control monkeys were consistently superior to the other groups and the isolate monkeys were superior to the enriched monkeys after the first 600 trials.

Unweighted means analysis of variance with rearing conditions (isolate-control-enriched) as the independent variable and delays (0- and 5-sec) and blocks as repeated measures yielded no significant effects of rearing conditions. However, a highly significant difference between delays (0-sec vs 5-sec) and between blocks (learning) did exist. The Fisher l.s.d. statistic revealed superior performance beyond the 0.05 level by the control condition over both other conditions at both levels of delays.

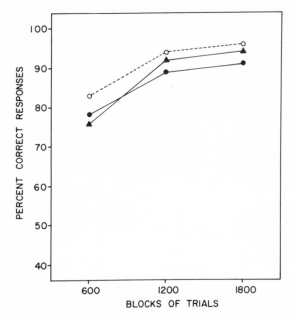

FIG. 17. 0-5 sec delayed response combined by isolated, enriched, and control monkeys. (▲) 8 isolated, 1085 days test age; (●) 12 enriched, 2910 days test age; (○) 17 control, 696 days test age.

Multiple Delayed Responses

The performances of the control, isolate, and enriched monkeys, with one incomplete isolate group omitted, are given in Fig. 18, and the data showed that the control groups were superior to the enriched groups and the completed isolate groups. These data are indeed surprising since there is no consistent order between deprivation and enrichment of early environment and adequacy of performance.

However, an unweighted means analysis of variance with the three rearing conditions as the independent variable and delays and blocks as repeated measures failed to disclose significant differences ($F = 2.32$, $df = 2/34$, $p < 0.10$). As in the case of the shorter delays, both the delays and blocks effects were highly significant (for delays $F = 149.3$, $df = 1/34$, $p < 0.01$, and for blocks $F = 9.3$, $df = 5/170$, $p < 0.01$). Furthermore, the Fisher l.s.d. statistic showed that all rearing conditions (isolate-control-enriched) performed significantly better on the two short delays combined than on the two long-delay levels combined ($p < 0.05$).

Oddity Learning Set

The data of Fig. 19 indicate little or no differences between the control and isolate conditions, but reveal a remarkably high competence level by the only enriched group (E-1) which has completed testing. These data are particularly surprising since E-1 is the youngest of all the groups plotted and is a group with an unsatisfactory adaptation record.

An unweighted means two-way analysis of variance with the three rearing conditions as an independent variable and learning (blocks) as a repeated measure revealed a significant difference for conditions ($F = 11.0$, $df = 2/14$, $p <$ 0.01) and for blocks ($F = 32.48$, $df = 1/14$, $p < 0.01$). The Fisher l.s.d. statistic was subsequently employed and indicated that the enriched group was superior to both the isolate and control groups ($p < 0.05$).

DISCUSSION

The overall data on discrimination, learning set, short and long delays, and oddity learning set performance of monkeys provide strong presumptive evi-

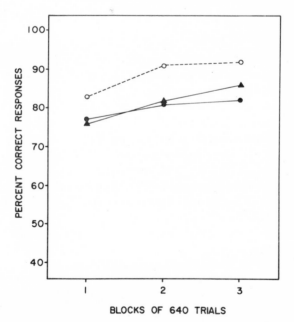

FIG. 18. Multiple delayed response performance by isolated, enriched, and control monkeys. (▲) 8 isolated; 1225 days test age; (●) 12 enriched, 436 days test age; (○) 17 control, 828 days test age.

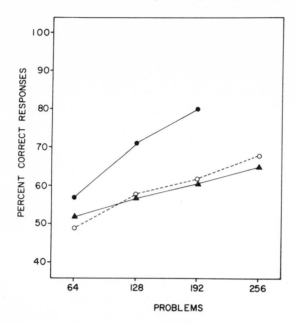

FIG. 19. Oddity learning set by isolated, enriched, and control monkeys (trials 1–6). (▲) 8 isolated, 1313 days test age; (●) 4 enriched, 767 days test age; (○) 17 control, 1070 days test age.

dence that learning or intellectual capability is neither enhanced by rearing in a socially enriched environment nor damaged by rearing in a drastically deprived environment. One deprived environment, early total social isolation, which subsequently destroys social, sexual, and maternal behaviors, leaves the monkey mind crystal clear and cognizant.

On two of the five learning measures tested in the present study, discrimination learning and multiple delays, there were no substantially significant differences among groups. On two tests, learning set and combined 0- and 5-sec delays, the enriched monkeys were significantly inferior to both other groups. On just one test, the oddity test, did the enriched monkeys surpass the other two groups in performance.

It is unfortunate that at the present time only one of the three enriched groups has completed oddity learning set. The performance of this group is remarkable in comparison with the performances of the animals in the other two rearing conditions and the performances made by equal-aged monkeys on the same test in other researches (Harlow *et al.* 1970). If the remaining two groups of enriched monkeys perform equally well, the data will be described and conceivably explained in a subsequent paper.

However, the learning data presently obtained from this battery of tests suggest no obvious performance differences by the monkeys raised under socially enriched conditions, long-term total isolation conditions, and control conditions. In other words, the long-term effects of enriched, as contrasted with deprived, environments are enormous in terms of social-emotional variables, but trifling or nonexistent when measuring learning or intellectual variables. Therefore, it is a grave mistake to confound assessment of learning ability with social and emotional variables.

We recognize that these conclusions are not in keeping with the data of previous studies which report striking effects of early deprived and enriched environments on learning performance of rats and dogs. It would be simple to dismiss the other data on the basis that man is a primate, and that rats and dogs are obviously lower animal forms.

Essential Considerations for Measuring Early Environmental Effects on Learning

We believe that unless all emotional factors which adversely affect learning have been eliminated or equated for controlled, deprived, and enriched animal groups, it is simply impossible to test the effect of good, bad, or indifferent environmental experiences on learning.

Importance of Prolonged and Extensive Adaptation

No animal, regardless of rearing condition, should be trained until all animals in all groups demonstrate equal adaptation to all of the learning test situations. This was the goal attempted in the present study, in which 20 to more than 100 adaptation days were allotted to our single simple test situation. However, it is doubtful if even these efforts achieved this end. Few investigators in this area have ever devised a comprehensive learning battery dependent on a single test situation. The task of adapting to multiple situations staggers the imagination. Some previous investigators apparently had no imagination to stagger.

After adaptation testing, all of our groups of monkeys attained respectable and responsible scores on our oddity learning-set problems, which are in fact insolvable problems for rats and probably for beagles. It appeared that adaptation testing was completely adequate. However, when the same monkeys were tested on short delayed responses, which we know from experience on multiple earlier studies to be emotionally taxing, the effect of relatively minor differences in early experience flared forth.

The enriched environment group 1 made scores so inadequate on the short delayed responses as to suggest that its members had suffered from bilateral frontal lobectomy. Actually, during the adaptation process the subjects in this group had been lured from the security of friends and family in the nuclear test situation (see Figs. 9 and 10), and they were apparently full of latent insecurity and hostility. During the delayed-response testing intervals these infant monkeys threatened the tester and ignored the problem. Fortunately these monkeys finally adapted after a thousand trials and made exemplary scores on subsequent multiple delay problems.

Similarly, isolate group I-3 was the slowest of all groups to adapt to the WGTA, but eventually made a respectable discrimination. However, the trauma of even short delayed-response testing resulted in totally inadequate learned performance until the completion of 1200 trials. Had we run an "abbreviated" 600-trial test we would have concluded that deprived environments produced idiocy. It would not have been the animals' idiocy, but the experimenters', and this holds for many or most antecedent studies.

Avoidance of Pain or Emotional Shock

An absolute requirement for measuring the effect of differential early environments on the learning capabilities of animals is that the tests themselves do not involve extraneous pain or fear-producing properties. Social isolation is not a state of organic nothingness; social isolation commonly leaves animals extremely hypersensitive to pain or fear, and this can hopelessly confound any subsequent assessment of learning ability. Assaying the effects of social deprivation or enrichment on emotional stability or instability is a very different function than that of measuring learning capability. It has long been recognized that learning variables and performance variables must be differentiated. Apparently, it has also occasionally been forgotten.

Any postisolation test involving pain or shock as an unconditioned stimulus is absolutely contraindicated. Monkeys, particularly socially isolated monkeys, freeze when shocked, and it is impossible to measure the I.Q. of an iceberg (Rowland, 1964). The confused literature on dogs at least indicates that social isolation produces abnormal responses to pain (Melzack & Scott, 1957). Dogs apparently suffer from neither fire nor desire.

Avoidance of Emotionally Disturbing Learning Tests

Another learning test which appears to be absolutely contraindicated as a measure for assaying the effects of deprived environments on learning is one which involves an open field or any variant of an open field. These are effective measures of emotionality in rats (Broadhurst & Levine, 1963; Ottinger & Denenberg, 1963), but emotionality must be ruled out to measure deprivation effects on learning per se. Tests such as the Hebb-Williams are partial variants of an open field situation and performance on them must be greatly influenced by emotional difficulties. An apparatus which lies in an open field that must be crossed before the problem can be faced becomes an instrument designed to test terror first, and mental manipulations second.

We are not trying to belabor any particular test or test situation, but data from any learning test involving strong emotional components are likely to give results either meaningless or directly contrary to the truth in assaying the effect of deprived environments on subsequent learning.

Equal and Adequate Adaptation of All Groups

If the learning capabilities of animals raised in deprived, normal, and enriched environments are to be compared, all subjects must be equally adapted to the testing situation or situations, assuming that this is possible. It is absolutely essential that the testing situation does not involve emotional trauma beyond that eliminated by the adaptation situation.

Adapting rats to a straight alley and then testing them on maze learning and reversal is undoubtedly proper, but adapting animals to a runway situation and then testing terriers (Thompson & Heron, 1954), or badgering beagles (Lessac & Solomon, 1969) on open field tests, barrier tests, shock avoidance, and pain threshhold situations proves nothing beyond the hopeless naivety of the experimenter. Of course, the prime example of inadequate adaptation is no adaptation at all.

Avoidance of Superficial or Cursory Learning Tests

If the effects of early experience are to be adequately assayed, a comprehensive battery of tests ranging in difficulty from problems which can be solved by infants to problems which challenge the intellectual capacities of the mature organism must be employed. The use of a single discrimination problem, a single discrimination reversal problem, or 100 delayed-response trials is devoid of meaning. Judging from our own experience, it is certain that the use of a small number of discrimination problems, discrimination reversal problems, and delayed-response trials is inadequate. Our own test battery requires 18 months to complete, and had we used an abbreviated battery the results would have been questionable.

If one is going to assay the effect of early experience on learning and intelligence, one should not grasp blindly at available tests, but should construct adequate test sequences before beginning the experiment. The fact that this may take 10 to 20 years should not deter the serious investigator.

There are two reasons why a comprehensive test battery is needed. The lengthy battery enables the animal to adapt to specific test situations if the previously formal adaptation sequences are in fact inadequate. A comprehensive battery enables one to assess specific kinds of intellectual impairment which may have occurred, if by any strange chance intellectual impairment was actually achieved.

Equal Experience on Component Parts of the Learning Problems

If one plans to assess the effect of two different environments, such as enriched and restricted, on learning, it is essential that the animals have equal experience with stimulus configurations which would directly transfer to learning in the postenvironmental training situation. If subjects are to be maze tested, they should not be given Hebb-Williams apparatus familiarization and maze familiarization in the enriched environment and none in the deprived before testing in a multiple discrimination apparatus (Krech et al., 1962). If groups of subjects are to be tested on open field tests or on tests placed in open fields, the enriched group should not be beagles taken for frequent walks about the laboratories, the campus, and from home to home (Lessac & Solomon, 1969). No techniques could more certainly insure better emotional adaptation by the enriched group and considerable specific learning transfer.

We present these injunctions not in anger, but in sadness. There now exists a wealth of intellectually poverty-stricken literature which shows that any knowledgeable experimenter who wishes to demonstrate that mammals raised in enriched environments are intellectually superior to those raised in deprived environments can achieve this goal. By conforming to simple fundamental laws of human stupidity this is easily achieved by some investigators using simple experimental designs and by others using experimental designs that they conceive to be recondite (Lessac & Solomon, 1969). The cheerless thought is that the experiments were created by human beings themselves reared in enriched environments, proving only that enriched environments alone are not adequate to facilitate thinking.

CONCLUSIONS

The results of these experiments raise serious doubts that differences in early, preadolescent environments leave any long-term effect on learning or intellectual capabilities. On only one measure was a group of environmentally enriched monkeys superior to monkeys raised in either moderately deprived or extremely deprived conditions, and on the other measures the enriched monkeys were frequently inferior to those suffering extreme social deprivation. The testing of all groups under the deprivation, control, and enriched conditions is not complete, and time must ensue before a completely definitive statement can be made.

Extremely comprehensive and prolonged measures of the social and emotional changes resulting from social isolation have long been made on our monkeys, and as a result we were aware of the violent and catastrophic personality disorders which we would encounter when animals raised for extended periods of time in a deprived environment were brought to a learning test.

Recognizing this, we designed our test situation so that effective or relatively effective adaptation to the single test situation could be achieved. We instituted a program of prolonged, self-paced adaptation to the test situation for the monkeys in all groups and all conditions. There was reason to believe that equal adaptation by all groups was approximately achieved.

For a variety of reasons, earlier investigators did not see this emotional problem or did not make adequate provision for it, either in the test situations or in the method of test adaptation.

It is our belief that the previous experimenters were wrong in assuming that social deprivation debased subsequent learning ability or that social enrichment enhanced subsequent learning ability. As of the present time the only conclusion that we can draw is that early environments greatly alter emotional and personality variables but have little or no effect on learning or intellectual variables.

REFERENCES

Alexander, B. K. The effects of early peer deprivation on juvenile behavior of rhesus monkeys. Unpublished doctoral dissertation, University of Wisconsin, 1966.

Bettelheim, B. Feral children and autistic children. *American Journal of Sociology*, 1959, **64**, 455-475.

Broadhurst, P. L., & Levine, S. Behavioral constancy in strains of rats selectively bred for emotional elimination. *British Journal of Psychology*, 1963, **54**, 121-125.

Chamove, A. S. The effects of varying infant peer experience on social behavior in the rhesus monkey. Unpublished master's thesis, University of Wisconsin, 1966.

Foley, J. P., Jr. The "baboon boy" of South Africa. *American Journal of Psychology*, 1940, **53**, 128-133. (a)

Foley, J. P., Jr. A further note on the "baboon boy" of South Africa. *Journal of Psychology*, 1940, **10**, 323-326. (b)

Griffin, G. A., & Harlow, H. F. Effects of three months of total social deprivation on social adjustment and learning in the rhesus monkey. *Child Development*, 1966, **32**, 533-547.

Hansen, E. W. Development of maternal and infant behavior in the rhesus monkey. Unpublished doctoral dissertation, University of Wisconsin, 1962.

Harlow, H. F. The development of learning in the rhesus monkey. *American Scientist*, 1959, **47**, 459-479.

Harlow, H. F. Age-mate or peer affectional system. In D. S. Lehrman, R. A. Hinde, & E. Shaw (Eds.), *Advances in the study of behavior.* Vol. 2. New York: Academic Press, 1969.

Harlow, H. F., Blomquist, A. J., Thompson, C. I., Schiltz, K. A., & Harlow, M. K. Effects of induction age and size of frontal lobe lesion on learning in rhesus monkeys. In R. L. Isaacson (Ed.), *The neuropsychology of development: A symposium.* New York: Wiley, 1968.

Harlow, H. F., & Harlow, M. K. Social deprivation in monkeys. *Scientific American*, 1962, **207**, 137-146.

Harlow, H. F., Harlow, M. K., Rueping, R. R., & Mason, W. A. Performance of infant rhesus monkeys on discrimination learning, delayed response, and discrimination learning sets. *Journal of Comparative and Physiological Psychology*, 1960, **53**, 113-121.

Harlow, H. F., Rowland, G. L., & Griffin, G. A. The effect of total social deprivation on the development of monkey behavior. *Psychiatric Research Reports*, 1964, **19**, 116-135.

Harlow, H. F., Schiltz, K. A., & Harlow, M. K. Effects of social isolation on the learning performance of rhesus monkeys. *Proceedings of the Second International Congress of Primatology*, 1969, **1**, 178-185.

Harlow, H. F., Thompson, C. I., Blomquist, A. J., & Schiltz, K. A. Learning in rhesus monkeys after varying amounts of prefrontal lobe destruction during infancy and adolescence. *Brain Research*, 1970, **18**, 343-353.

Itard, J.-M.-G. *The wild boy of Aveyron.* New York: Century, 1932.

Krech, D., Rosenzweig, M. R., & Bennett, E. L. Relations between brain chemistry and problem-solving among rats raised in enriched and impoverished environments. *Journal of Comparative and Physiological Psychology*, 1962, **55**, 801-807.

Lessac, M. S., & Solomon, R. L. Effects of early isolation on the later adaptive behavior of beagles. *Developmental Psychology*, 1969, **1**, 14-25.

Lorenz, K. Der Kumpan in der Umwelt des Vogels. *Journal für Ornithologie, Leipzig*, 1935, **53**, le7-2, 289-413.

Melzack, R., & Scott, T. H. The effects of early experience on the response to pain. *Journal of Comparative and Physiological Psychology*, 1957, **50**, 155-161.

Mitchell, G. D., Raymond, E. J., Ruppenthal, G. C., & Harlow, H. F. Long-term effects of total social isolation upon behavior of rhesus monkeys. *Psychological Reports*, 1966, **18**, 567-580.

Ogburn, W. F. The wolfboy of Agaro. *American Journal of Sociology*, 1959, **64**, 449-454.

Ogburn, W. F., & Bose, N. K. On the trail of the wolf-children. *Genetic Psychology Monographs*, 1959, **60**, 117-193.

Ottinger, D. R., Denenberg, V. H., & Stephens, M. K. Maternal emotionality, multiple mothering, and emotionality in maturity. *Journal of Comparative and Physiological Psychology*, 1963, **56**, 313-317.

Rosenblum, L. A. The development of social behavior in the rhesus monkey. Unpublished doctoral dissertation, University of Wisconsin, 1961.

Rowland, G. L. The effects of total social isolation upon the behavior of rhesus monkeys. Unpublished doctoral dissertation, University of Wisconsin, 1964.

Sackett, G. P. Monkeys reared in visual isolation with pictures as visual input: Evidence for an innate releasing mechanism. *Science*, 1966, **154**, 1468-1472.

Singh, J. A. L., & Zingg, R. M. *Wolf children and feral man.* New York: Harper, 1942.

Singh, S. D. Urban monkeys. *Scientific American*, 1969, **221**, 108-115.

Spitz, R. A. Hospitalism: An inquiry into the genesis of psychiatric conditions in early childhood. *Psychoanalytic Study of the Child*, 1945, **1**, 53-74.

Spitz, R. A. Anaclitic depression. *Psychoanalytic Study of the Child*, 1946, **2**, 313-342.

Squires, P. C. Wolf children of India. *American Journal of Psychology*, 1927, **37**, 313-315.

Thompson, W. R., & Heron, W. The effects of restricting early experience on the problem-solving capacity of dogs. *Canadian Journal of Psychology*, 1954, **8**, 17-31.

Wasserman, J. *Kasper Hauser.* New York: Liveright, 1928.

CHAPTER 7

Some Differences between Human and other Primate Brains[1]

Norman Geschwind

Most of the papers in this book have been devoted to the capacities of nonhuman primates; but some of the authors, particularly Weiskrantz and Premack, have been concerned in part in their studies with searching for continuities with human behavior. Like them, I too suspect that the apparently unique capabilities of man did not suddenly appear without precedent in the course of evolution. The possibility must be entertained, however, that there may be very great gaps between man and the other primates since it is conceivable that some of the intermediate stages may have died off. I can only point out here some features of the human brain that appear to be special.

One of the most striking features of the human brain is dominance. Whatever language is, one of its most striking features is that in the adult human brain, the capacity either to produce it or understand it is overwhelmingly on the left side. The degree of left-sided predominance shows up clearly in some facts of clinical observation. Out of 100 right-handed patients with lesions of the right hemisphere, and indeed the grossest of lesions, it would be surprising to find more than one with any degree of language disorder. On the other hand, of 100 similar patients with lesions of the left hemisphere, about 80% will suffer from aphasia. Furthermore, many will remain permanently and severely aphasic as the result of a very small lesion in an appropriate area of the left hemisphere which causes no elementary neurological deficit.

The contrast between the capacities of the two hemispheres perhaps can be made more dramatic if we compare language abilities of the adult right hemisphere with the linguistic abilities of Dr. Premack's chimpanzee. Many patients

[1]Some of the work reported here was supported in part by Grant NS 06209 to the Boston University School of Medicine.

with limited left-hemisphere damage and completely intact right hemispheres have much less ability to comprehend or express symbolic material than does the chimpanzee trained by Dr. Premack.[2]

One should not, of course, assume that the human right hemisphere has less *potential* capacity for language than the chimpanzee brain. If the left hemisphere is damaged in childhood the right hemisphere will acquire language. Furthermore in occasional adults the right hemisphere shows a very significant ability to take over language functions. On the other hand, in most cases this ability is either limited, as in the case of Smith (1966) on whom Burklund had carried out a left hemispherectomy for tumor, or absent. This can be illustrated in another way. The syndrome of pure alexia without agraphia, in which the patient loses the ability to read, but retains all other speech functions was already well-recognized before 1890, but the first postmortem examination was published by Dejerine (1892). There was destruction in his patient of the left visual cortex. Visual input therefore could reach only the right visual region. The patient had a second lesion in the splenium of the corpus callosum. As a result of this lesion the visual input could not be transmitted to the speech regions of the left hemisphere. This syndrome was, in fact, the first proven demonstration of the importance of the corpus callosum in transferring information between the hemispheres. These pathological findings have been confirmed repeatedly, e.g., in the case of Geschwind and Fusillo (1966). It is important to realize that as a result of this very limited callosal lesion the right hemisphere which is separated from the speech regions on the left loses its ability to comprehend written language, whether tested by verbal response, or by matching tasks (e.g., matching a picture to a word) using either the right or the left hand. In several of the cases of Gazzaniga, Bogen, and Sperry (1965) patients who had undergone total section of the corpus callosum and anterior commissure could do word-picture matching with the right hemisphere. These cases, however, were all patients who had had long-standing epilepsy dating from early life, and illustrate the point that early brain lesions often lead to greater language capacities in the right hemisphere than would normally be found. The extreme example of this is, of course, the condition of agenesis of the corpus callosum.

Is there any known example of cerebral dominance in mammals below man? At the present writing, none has, to my knowledge, been demonstrated. There are experiments in which unilateral lesions lead, for example, to marked inattention to the opposite side, as in the experiments of Welch and Stuteville (1958) in the macaque. There was no dominance demonstrated since the effects of lesions on either side were apparently of equal magnitude. More recently Weiskrantz[3] has reported experiments in which unilateral lesions produce bilateral effects. This important finding, however, will not necessarily be a demonstration of

[2]See Chapter 3, this volume.

dominance if it turns out that the impairment produced is equally severe from either side. Weiskrantz's experiments may, however, be relevant to the situation in human left-handers in whom there is evidence that a unilateral lesion on *either* side will produce aphasia (Gloning, Gloning, Haub, & Quatember, 1969) thus suggesting that dominance on the whole is not as well developed in this group as in a group of right-handers.

It thus appears fair to say that at this time, no evidence of dominance has been found in mammals below man. What would be required would be an experiment in which a lesion on one side produced an effect that could not be elicited by a lesion in the symmetrical region on the other side. It is, of course, conceivable that many subhuman mammals do normally use one side of the brain predominantly for certain functions, but compensate rapidly with the other side after unilateral lesions. The classical ablation experiment might not be adequate to reveal this.

It is interesting that in submammalian forms there are examples of dominance, e.g., in birds (Nottebohm, 1970). These brains are, however, so remote from that of man that it is conceivable that they represent not an earlier stage of the development of human dominance, but a separate development.

What underlies human cerebral dominance for language? It has generally been stated in the literature (e.g., Bonin, 1962) that anatomical differences could not explain dominance, and that it must therefore by the result of subtle physiological differences. My colleague, Dr. Walter Levitsky, and I became interested in this problem, and following up earlier leads, we were able to show that, contrary to usual opinion, the human brain had gross anatomical asymmetries in areas known to be involved in speech (Geschwind & Levitsky, 1968). The area studied was the planum temporale, which lies on the upper surface of the temporal lobe, bordered anteriorly by Heschl's gyrus and in back by the posterior border of the sylvian fossa. This area is larger on the left in 65% of brains, larger on the right in 11%, and equal in 24%. The right-left difference is $p < 0.001$. The outer border of the planum averages 3.6 ± 1.0 cm in length on the left, 2.7 ± 1.2 cm on the right ($p < 0.001$). The left planum is therefore on the average 0.9 cm or 33 1/3% longer than the one on the right. This area, which is greater on the left, is clearly, as can be seen from the cytoarchitectonic studies of Economo and Horn (1930), auditory association cortex, and is obviously the extension on the superior surface of the temporal lobe of Wernicke's area (which lies on the convexity in the posterior portion of the superior temporal gyrus), which has long been known to be one of the most important areas involved in speech.

More recently Wada (1969) has confirmed our findings in the brains of adults, but has also demonstrated that these asymmetries are present in the brains of newborn infants. It seems likely at this time that this asymmetry probably is

[3] Personal communication to the author.

unique to man, and is not present in the brains of subhuman primates, at least as far as present knowledge is concerned.

This predominantly unilateral organization of the human brain for language functions is also reflected in the memory system. Let us consider again the patient of Geschwind and Fusillo (1966) mentioned earlier. In addition to his right hemianopia and alexia which were persistent over 9 months (until his death) he also showed after his stroke, an amnesic syndrome which lasted for about 3 months and then cleared. How do we account for this transient amnesic syndrome? At postmortem the brain showed a clear area of destruction of the left hippocampal region with associated degeneration of the left fornix. It seems likely that we must attribute the transient memory loss to the left hippocampal lesion. It is highly unlikely that the right hippocampal region had been ischemic without infarction and had recovered, since 3 months is too long for such recovery. It is more likely that the memory disorder resulted from destruction of the hippocampal region normally used by the speech regions, and that it took about 3 months for the longer, alternative routes to the right hippocampal region to come into play.

As further evidence, I have seen other cases with right hemianopia and alexia who have shown a similar disorder of memory. The association of these signs is explained by the fact that all the relevant anatomical structures are within the distribution of the posterior cerebral artery, occlusion of which may produce all the above effects. Dr. C. Miller Fisher has told me in a personal communication that he has seen several cases of transient memory disorder after unilateral lesions, and these have almost invariably been on the left.

There may appear at first to be some discrepancy between the dramatic memory disorder in these cases and the much milder verbal memory disorders which have been brought out by the elegant studies of Milner (1962) in patients who had undergone left anterior temporal lobectomy. It should be pointed out that these were all patients who had been suffering from epilepsy for some time, thus disrupting the normal functioning of the left hippocampal region, and had probably made the same adjustments over a long time that the patients with sudden hippocampal destruction made over a shorter period. It is therefore understandable that the disorders in the surgical cases are milder.

The special role of the left hippocampal region in memory for language is also brought out in the case of Geschwind, Quadfasel, and Segarra (1968). This patient had suffered from carbon-monoxide poisoning which she survived for 9 years. During this period she never uttered a sentence of propositional speech and never showed any evidence of comprehension of language. On the other hand, she repeated perfectly, without articulatory disorder, sentences spoken to her by the examiner. Even more dramatic was her ability to carry on verbal learning. A record of a song which had not existed before her illness was played to her several times. She would, after a few trials, start to sing along with the record. Then the record would be started and turned off as soon as the patient began to sing. She would then continue singing the song correctly, both words and music, to the end.

At postmortem there was widespread damage in the brain. What was important however, was not so much what was damaged, but what was spared. Preserved were the classical speech regions, including Wernicke's area and Broca's area and the connections between them, as well as the auditory input and the motor output pathways from these regions. The speech system was thus isolated from the rest of the cortex. In addition, however, the hippocampal region was spared. It seems reasonable that in this patient what was preserved was the system necessary for carrying on verbal learning, i.e., the classical speech regions and the medial temporal region.

It seems likely that the remarkable linguistic abilities shown by chimpanzees, as in the work of Dr. Premack[4] and Gardner and Gardner (1969), would not be shared by the macaque. It would be my guess that this is probably related to an increased development of the angular gyrus region in the chimpanzee as compared with the macaque. It is, in fact, this region of cortex which has expanded most markedly in man compared to the other primates. I have discussed elsewhere (Geschwind, 1964, 1965) the reasons why this area may be so essential to the development of language in the human sense.

REFERENCES

Bonin, G. V. Anatomical asymmetries of the cerebral hemispheres. In V. B. Mountcastle (Ed.), *Interhemispheric relations and cerebral dominance.* Baltimore: Johns Hopkins Univ. Press, 1962.

Dejerine, J. Contribution à l'étude anatomo-pathologique et clinique des différentes variétés de cécité verbale. *Mémoires de la Société de Biologie,* 1892, **4,** 61-90.

Economo, C. V., & Horn, L. Uber Windungsrelief, Masse und Rindenarchitektonik der Supratemporalfläche. *Zentralblatt für die Gesamte Neurologie und Psychiatrie,* 1930, **130,** 678-757.

Gardner, R. A., & Gardner, B. T. Teaching sign language to a chimpanzee. *Science,* 1969, **165,** 664-672.

Gazzaniga, M. S., Bogen, J. E., & Sperry, R. W. Observations on visual perception after disconnexion of the cerebral hemispheres in man. *Brain,* 1965, **88,** 221-236.

Geschwind, N. The development of the brain and the evolution of language. In C. I. J. M. Stuart (Ed.), *Monograph series on languages and linguistics,* No. 17. Washington, D.C.: Georgetown University Press, 1964.

Geschwind, N., Disconnexion syndromes in animals and man. *Brain,* 1965, **88,** 237-294; 585-644.

Geschwind, N., & Fusillo, M. Color-naming defects in association with alexia. *Archives of Neurology,* 1966, **15,** 137-146.

---

[4] See Chapter 3, this volume.

Geschwind, N., & Levitsky, W. Human brain: Left-right asymmetries in temporal speech region. *Science*, 1968, **161**, 186-187.

Geschwind, N., Quadfasel, F. A., & Segarra, J. M. Isolation of the speech area. *Neuropsychologia*, 1968, **6**, 327-340.

Gloning, I., Gloning, K., Haub, G., & Quatember, R. Comparison of verbal behavior in right-handed and nonright-handed patients with anatomically verified lesion of one hemisphere. *Cortex*, 1969, **5**, 43-52.

Milner, B. Laterality effects in audition. In V. B. Mountcastle (Ed.), *Interhemispheric relations and cerebral dominance*. Baltimore: Johns Hopkins Univ. Press, 1962.

Nottebohm, F. Ontogeny of bird song. *Science*, 1970, **167**, 950-956.

Smith, A. Speech and other functions after left (dominant) hemispherectomy. *Journal of Neurology, Neurosurgery & Psychiatry*, 1966, **29**, 467-471.

Wada, J. Interhemispheric sharing and shift of cerebral speech function. Paper presented at the 9th International Congress of Neurology, New York, September 1969.

Welch, K., & Stuteville, P. Experimental production of unilateral neglect in monkeys. *Brain*, 1958, **81**, 341-347.

CHAPTER 8

Similarities in the Cognitive Processes of Monkeys and Man[1]

Lee W. Gregg

The goal most of us share is to explain the behavior of organisms in terms of the processes and functions of their neurophysiological systems. Implicitly, we assume that the structures and organizations of these systems are alike in some respects; and further that our descriptive concepts for saying what we mean by behavior comprises a reasonably complete lexicon. Obviously there are common features from rat brain to monkey to man and such gross behavioral concepts as motivation, perception, and learning have been useful chapter headings for a very long time.

This work on comparative aspects of nonhuman primate cognition is, therefore, a matter of the degree to which we wish to disagree. Hopefully my contribution will be to show that cognitive processes, both in monkey and man, have certain fundamental properties in common that can be described by a few elementary information processes. The implication is that similar brain structures may therefore be present. Of course, the same unsolved problems of human cognition must surely remain unsolved for nonhuman primates, and what we know least about will generate the greatest opportunities for disagreement.

The set of elementary information processes we have found to be critical for explaining human cognitive behavior are a subset of the ones included in Information Processing Language V (IPL-V) (Newell, 1961). Their definition is in terms of operations a computer can perform and they have been used in constructing a variety of computer simulation models. Most of you are familiar with the notion that a computer model takes the form of a program where data structures represent the perceptual relations of environmental objects and events

[1] This research was supported by a grant from the Public Health Service, National Institute of Mental Health, Research Grant MH-07722.

as well as organizations of potential sequences of action or tests that the human learner or problem-solver can carry out. These data structures define the organization of human memory. Process descriptions are expressed as executable instructions governing the flow of information transformations on such data structures.

Very early we recognized the need for a suitable language for programming computers. IPL-V was invented to incorporate the symbolic manipulations necessary to implement computer-simulation models. Many of the features of the language deal with housekeeping or other tasks specific to using the computer effectively. But the critical features were the "primitives" that were considered to be important then and necessary now for capturing the symbolic manipulations of human thought.

Briefly, these primitives are: create a symbol, test if symbol A is identical with symbol B, find the value V of an attribute A of an object named by symbol C, and assign a symbol V as value of attribute A on object C. A system using these constructs must, additionally, communicate the sense of the outcome of the test operation and have the capacity for locating and storing symbols. Symbols may designate locations or internal names, or they may have referents in the usual sense. Any object or event may be named by a symbol. Any stimulus, as long as it generates sensory events, can have a symbolic name.

Out of these few elementary information processing elements we have constructed models of concept attainment, verbal learning, and a variety of problem-solving behaviors from chess play to simple guessing games. We have not tackled the more formidable, but potentially "lower level," mental processing of subnuman species. However, a starting point is suggested by Professor Meyer's statement that "monkeys learn concepts." Are monkey concepts the same as human ones?

CONCEPTS AND HABITS

Meyer[2] draws a distinction between concepts and habits based on the essential continuity of performance during acquisition of new problems while monkeys undergo acquisition and extinction of the reversal learning set. He ascribes intraproblem performance to a collection of mechanisms that appear to be quite stable. Specific-habit learning of novel object-quality discriminations proceeds in a relatively independent fashion. Qualitatively, the effect of concept learning, the formation of learning sets, is to introduce sharp discontinuities in performance when experimental conditions are changed. Although one would like to have a quantitative statement about the learning rates for the separate processes, it is probably the case that no direct comparisons, that make sense, can ever be made because, as Meyer concludes, the intraproblem learning is probably governed by more than one mechanism. What precisely could these "microprocesses" be?

I submit that monkeys learning object-quality discriminations must be creating internal representations of the object pairs in much the same way that humans familiarize nonsense syllables in a verbal learning task. Any discriminative performance, above chance levels, requires at least the capacity to sort an object on the basis of tests of, say, its visual-spatial properties. We have chosen to represent the outcome of this process as a sorting tree or discrimination net in computer simulations of human verbal learning (Feigenbaum and Simon, 1962; Simon and Feigenbaum, 1964).

One program, Elementary Perceiver and Recognizer (EPAM), identifies a collection of microprocesses sufficient for the job. The discrimination tree grows as the familiarization process creates new tests at nodes. At a node values are assigned to the test attribute and lead finally to a "terminal node." At the terminal, a recognizable "image" of an object can be found. Associated with the image may be additional information or description; for example, the fact that the object was previously the rewarded one.

A mechanism at least this complicated is necessary for an organism to exhibit discrimination learning. To explain the positive-transfer effect that Bettinger[2] found for the repeated circle and triangle problem we must suppose that the information associated with the image or memory trace of one or both objects is the reward value or its correctness from prior problem presentations. But for any organism that forms a learning set, this means that an organization or reorganization of central nervous system functions must provide a mapping of values along a dimension. Cognitive control of behavior therefore comes about because internal events—the idea of the name of the dimension—direct attention to some but not all information contained in the object so that stimulus values generate a unique "switching" as a function of internal states.

Concepts as used by Meyer produce changes in levels of performance when the concept, e.g., reversal set, is no longer appropriate, as in the reversal-cue problem. The performance changes appear to be substantial, almost discontinuities rather than the slow incremental modifications of specific-habit learning. One interpretation of what a concept does for the learner is related to the control of attention. Having the reversal set is to switch attention from the previously correct object to the "other than focal" object of the pair. Hence the information available to the cognitive system changes. The monkey performing effectively under a reversal set disregards the descriptive information about the incorrect object and attends to the other object.

---

[2]See Chapter 4, this volume.

In human verbal learning, we can induce sets in many ways. But the most direct way is to instruct the subject in what to look for and attend to in the task. Most often subjects are told to respond to each new item in a serial or paired associate list. In the anticipation procedure, subjects are encouraged to respond. Of course, not all subjects do what they are told to do. A recent controversial issue in verbal learning was the question of all-or-none vs incremental learning raised by Rock (1957). Here the resolution of apparently contradictory results depended on simple differences in the subject's control of his own attention. To the extent that information was retained about items long enough for the subject to learn the paired nonsense syllables, performance changes over trials appears to be all-or-none (Gregg & Simon, 1967a). The concept of attending to each item as it is presented is lost on many subjects who decide on a one-at-a-time strategy for managing the surfeit of information imposed by the presentation of lists of items.

An information-processing analysis of concepts attempts to spell out the minimum requirements of data and processes that will generate conceptual behavior. The concepts that Meyer's monkeys have when they have formed a learning set appear to be perceptual or attention-directing rules like human sets (Haygood and Bourne, 1965). To perform effectively under the guidance or set, information must be available to call out the set from its representation in memory. At least one test must be made to distinguish among the different courses of action that may be taken, and there must be appropriate information, say of the relevant perceptual dimensions, on which the actions are based. This concept of a concept produces a fairly complicated program or subroutine. Examples of concepts modeled as subroutines are found in several of our simulations of conceptual behavior (Gregg, 1967; Gregg & Simon, 1967b).

For humans with language capabilities concepts have names. There is not much mystery about the way these complex subroutines are called; language symbols serve as retrieval cues. This cannot be the case for Meyer's subjects. Monkeys and apes cannot learn language—or can they?

LANGUAGE AND COGNITION

Premack's[3] functional analysis of language is an important contribution to our understanding of how cognitive acts control behavior. His answer to the question "What is language?" is a list of behaviors that we as language users can exhibit. His choice of items for that list is interesting. Although he disclaims any systematic or exhaustive basis for generating those items, his selection bears a remarkable resemblance to our list of elementary information processes.

The function "word" assigns a name. Therefore, a word symbol (blue triangle for apple) must create an internal symbolic element. A "sentence" is a list of words. Premack's functional test of sentence comprehension required Sarah the

[3] See Chapter 3, this volume.

chimp to execute an appropriate sequence of actions as in the sentence "green on red," placing a green card on a red card already on the table. A minimal program for this would require the execution of at least three subprograms—find green card, find red card, and "place on" with the two input symbols. Premack's sentence production test required Sarah to generate the sentence "green on red" by analyzing the environmental events carried out by the experimenter. In either case, the sentence represented by the vertical array of plastic chips on the board had to have an internal counterpart. More direct evidence for such internal representations is the features analysis of symbolically denoted fruit. The symbol "blue chip" stands for an object "apple" that has attribute values color—red; shape—circle. Clearly Sarah is capable of "finding the value V of attribute A on object C."

Although Premack states that most of his effort was directed toward mapping concepts that the chimp already knows, his training procedures were sufficient to show that Sarah could "*assign* the value (brown) to attribute (color of) of object (piece of chocolate)." The mapping procedures are based on the interrogative function and tell us that discriminative tests, e.g., "is *symbol* A different from *symbol* B," are within Sarah's capabilities.

What Premack calls "metalinguistics" is a necessary function of language that provides a level of indirection required to locate a symbol in a memory structure. Why is indirection necessary? On the surface, it would seem that Premack's training procedure simply establishes an association between two things. The blue plastic chip which is to be the "name of" an object is one. The object itself is the other. On closer examination, we see that the training procedure introduces a new element—actually two elements—namely, the plastic chips that are the language symbols for "name of" and "not the name of." Now in one sense the chimp already knows that the blue triangle symbolizes an apple. And so it seems almost unnecessary to introduce a special operation to verify this result. However, as Premack observes, the *concept* "name of" has a peculiar status.

Both the blue plastic chip and the object apple have internal names, i.e., internal representations however they may be realized in brain structures. To say that the blue chip stands for the object apple is to posit an associational relationship, a link, between two internal symbols. But the associative link cannot be isomorphic because then the chimp would be as likely to eat blue chips as apples. We can diagram the state of affairs as follows.

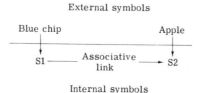

External symbols

Blue chip   Apple

S1 ———— Associative ———— S2
               link

Internal symbols

The internal symbols S1 and S2 must occupy different locations. What is the nature of the associative link? It might be a unidirectional pointer from S1 to S2. In this case we could call S1 the "location of" S2. But this would preclude associating any other information with S1; for example, the fact that the blue chip can cling to the word board or is itself inedible. To resolve the issue, it must be the case that an additional internal symbol is necessary. Associated with the symbol S1 must be a pointer to the location of S2, call it S2'.

The final step is the one achieved by Premack when he introduced the plastic word for "name of." To be able to distinguish the property of the blue chip having to do with apples rather than boards or eating activity, the attribute "name of" must take on the value S2'. Let S3 be the internal symbol for "name of."

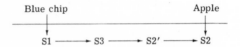

Notice that there is only one primed symbol in the diagram, S2'. In fact, each of the internal symbols must have their own locations, S1', S3' and so on. Without explicit location information the system could not operate, and we would be in Wonderland with Alice.

> "The *name of the song is called* 'Haddock's eyes'."
> "Oh, that's the name of the song, is it?" Alice said, trying to feel interested.
> "No, you don't understand," the knight said, looking a little vexed.
> "That's what the *name is called.* The *name really is* 'The aged aged man'."
> "Then I ought to have said, 'That's what the song is called'?" Alice corrected herself.
> "No, you oughtn't: that's quite another thing! The *song is called* 'Ways and means': but that's only what *it's called*, you know!"
> "Well, what is the song, then?" said Alice, who was by this time completely bewildered.
> "I was coming to that," the knight said. "The *song really is* 'A-sitting on a gate': and the tune's my own invention."

(From Lewis Carroll's *Alice in Wonderland,*
Through the Looking Glass, Chapter 8.)

In IPL-V, an appropriate structure for Carroll's fancy is provided by an organization of computer words with the following format.

Name	Symbol	Link
Haddock's eyes	The aged aged man	Ways and means
Ways and means	A-sitting on a gate	ϕ

At location "Haddock's eyes" we can find the symbol "the aged aged man" and so to say (or print in the case of the computer) the name of the song we must locate it via "Haddock's eyes" and execute an appropriate motor response with the symbol "The aged aged man" as input. The link through the symbol "ways and means" makes it possible to locate the song itself. Obviously, to *sing* the song one must execute a singing subroutine with "A-sitting on a gate" as input! The symbol ϕ in the link of the word called "ways and means" is an IPL convention to indicate the termination of the list structure.

Let me summarize by making two observations. First, the concept "name of" was not in fact a completely new concept "inculcated by the language training." Language learning must always map prior cognitive structures. There is a "perceptual language" defined by prior experience and organization of that experience by the cognitive system. Second, I believe that the demonstrations presented by Premack show that the chimpanzee is capable of acquiring language functions almost as complete as man. If true, this means that the cognitive system of that nonhuman primate species must be capable of organizing information in ways that are at least as complicated as suggested above.

There are, however, two important respects in which the differences between man and monkey may be very great. Both have to do with phenomena of short-term memory (STM), the subject of Dr. Jarrard's study. One problem is the high distractability of nonhuman primates which might imply marked differences in STM capacity. The magic number 7 shrinks in humans to about 2 or 3 when the information content requires place-keeping in the performance of cognitive tasks. Perhaps it could be shown that this already small value shrinks to one place-keeper or none at all in the nonhuman primate. The second potentially devastating difference is in the inability of the nonhuman primate to articulate speech sounds. Acoustic confusibility, a topic of recent interest (Conrad, 1964; Wickelgren, 1965) would be hard to observe if the monkey's STM subsystem did not encode sounds. More important, rehearsal as we know it might not be possible in species other than man.

SHORT-TERM MEMORY AND THE CONTROL OF ATTENTION

The data presented by Jarrard and Moise[4] show in a qualitative way similarities to measures of STM in human subjects. Jarrard has detailed the limitations: human data are based on verbal stimuli; human experiments most often measure recall of multiple-item lists; and human subjects can articulate and rehearse the stimulus materials. In other words, the delayed-matching-from-sample (DMS) task is too simple a task to make strong statements about STM as a theoretical construct or a neurophysiological entity.

What is DMS good for? I think it could be most useful in mapping perceptual units in much the same way Premack initiated his study of the chimp. By examining changes in the shape of the delay curves as a function of the number of repetitions of sample stimuli and total time of their presentation, it should be possible to trace the integration of novel, complex stimuli. It might be possible to discover the perceptual dimensions upon which familiarization proceeds. For human subjects our estimates of the time required to transfer a chunk from STM to LTM (long-term memory) is about 7 sec. Our definition of a chunk is "a highly familiar unit—a letter, digit, or common word." Simon (1970) asked the question "How big is a chunk?" and ran into trouble when trying to deal with stimuli other than simple, easily verbalized names. Perhaps, the DMS task could be used for both human and animal subjects. I wonder if the time constants would be the same. In the rehearsal study, Jarrard obtained statistically significant differences for 1, 2, and 4 repetitions of the simple colored light stimuli he used. But the absolute differences were not as large as one might expect. I think this is because of the very fast reaction time training given these animals. There was just not enough time for an animal without language symbols to store (in LTM) the attributes of the sample. What I am suggesting is that DMS does not measure STM in the monkey but rather the ability to create internal representations of familiar stimuli. For human subjects, the DMS task poses a somewhat different problem, related to STM function but only indirectly related. The subject's problem is to develop suitable retrieval cues or rehearsal strategies so that the sample item may be recognized at the termination of the delay interval.

Unfortunately, some important data were missing from Jarrard's program of research. He mentioned a study using the rehearsal paradigm but with a neutral, white light preceding the sample stimulus 1, 2, or 4 times. The hypothesis is that merely controlling attention to the locus of the sample, is sufficient to produce increases in retention comparable to the repetition of the sample itself. It appears to me that the issue of maximum disagreement in comparing cognitive functions across species is to determine whether attention processes or memory structures create the apparent gap between human and nonhuman levels of performance.

[4]See Chapter 1, this volume.

A behavioral approach we are just now starting is to map concepts of, say a chimp, in much the same way Premack interrogated Sarah. Then we will present novel tasks constructed to draw on the available conceptual resources of the ape. The question is "Can we communicate instructions about the problem to the chimp so that he can program his own solution behavior?" The difficult training problem we face in attempting to communicate the requirements of a novel task to a nonhuman primate might provide an approach to the solution of the problem of human program induction.

Finally, let me relate what we have learned so far working with a chimp named "Honey" at our local zoo. Miss Barbara Bessey contacted the trainer to arrange for our exploratory program in mapping concepts. Her first few sessions were conducted in the trainer's presence. Miss Bessey presented a simple detour problem to Honey, removing a lifesaver strung on a straight wire. The chimp was unrestrained, eyes darting about, hopping on and off a chair placed at the desk between the trainer Johnny and Miss Bessey. "First we must get her attention," said Johnny. He then hit Honey hard with a 12-in. piece of lead pipe. Honey worked at the task for about 5 sec . . . .

Monkeys have concepts, so do people. The chimpanzee can acquire concept names, as do humans. And a special kind of language facility makes it possible to communicate with a particular chimp. But as is so often the case, these statements about the nature of cognitive behavior rests on some special instances, particular conditions of observation. The most striking qualification is that Sarah's performance depended on sentences displayed in an external memory—the board on which plastic chips were placed.

Perhaps there are crucial differences in cognitive systems that are a function of the number of different memory structures available to them. The distinction between LTM and STM, the "many memory assumption," is well grounded in the last decade of research on human information processing. There may of course be three, four, or more memory subsystems similar to the Sperling (1960) very short-term store, or the intermediate-term memory that Sternberg (1969) scans. Memory subsystems may be modality specific or functionally delimited. Dr. Geschwind[5] suggests that an area of the cortex in the region of the angular gyrus may play an important role in language function. Perhaps that function is specific to establishing continuity of flow in the sense of place-keeping in human conversational ability.

REFERENCES

Conrad, R. Acoustic confusions in immediate memory. *British Journal of Psychology*, 1964, **55**, 75-84.
Feigenbaum, E. A., & Simon, H. A. A theory of the serial position effect. *British Journal of Psychology*, 1962, **53**, 307-320.

[5] See Chapter 7, this volume.

Gregg, L. W. Internal representations of sequential concepts. In B. Kleinmuntz (Ed.), *Concepts and the structure of memory*. New York: Wiley, 1967.

Gregg, L. W., & Simon, H. A. An information processing explanation of one trial and incremental learning. *Journal of Verbal Learning and Verbal Behavior*, 1967a, **6**, 780-787.

Gregg, L. W., & Simon, H. A. Process models and stochastic theories of simple concept formation. *Journal of Mathematical Psychology*, 1967b, **4**, 246-276.

Haygood, R. C., & Bourne, L. E. Jr. Attribute-and rule-learning aspects of conceptual behavior. *Psychological Review*, 1965, **72**, 175-195.

Newell, A. (Ed.) *Information processing language-V manual*. Englewood Cliffs, N. J.: Prentice-Hall, 1961.

Rock, I. The role of repetition in associative learning. *American Journal of Psychology*, 1957, **70**, 186-193.

Simon, H. A. How big is a chunk? Paper presented at the 41st annual meeting of the Eastern Psychological Association, Atlantic City, April 1970.

Simon, H. A., & Feigenbaum, E. An information processing theory of some effects of similarity, familiarization, and meaningfulness in verbal learning. *Journal of Verbal Learning and Verbal Behavior*, 1964, **3**, 385-396.

Sperling, G. The information available in brief visual presentations. *Psychological Monographs*, 1960, **74** (11, Whole No. 498).

Sternberg, S. Memory-scanning: Mental processes revealed by reaction time experiments. *American Scientist*, 1969, **57**, 421-457.

Wickelgren, W. A. Acoustic similarity and intrusion errors in short term memory. *Journal of Experimental Psychology*, 1965, **70**, 102-108.

CHAPTER 9

Species Differences in "Transmitting" Spatial Location Information[1]

Raymond C. Miles

The data that will be presented and my comments are most relevant to researchers concerned with memory functions. It was interesting to learn that Professor Weiskrantz[2] was able to deduce some commonality among the behaviors of rats, monkeys, and men following damage to the inferotemporal neocortical regions. His thoughtful analyses and alternative interpretation were instigated by some apparent discrepancies when the human and animal data were compared. In the animal subjects, the alleged "amnesic" memory deficit was either nonexistent or only very indirectly inferable. Both rats and monkeys demonstrated adequate recall of previous learning but shared the common symptom of being less able to inhibit alternative response tendencies.

In addition to the apparent differences between animal and human cases, Dr. Weiskrantz was clearly aware of the vagueness of the data and descriptions pertaining to the human "amnesic" syndrome. It is fortunate that the inconsistencies were not explained away but became the basis for further clarifying research.

The described experiments indicated that the disability could not be conveniently attributed to some short-term memory (STM) or long-term memory (LTM) dysfunction. Of more importance, a new conceptual orientation toward the baffling "amnesic" syndrome was suggested. The brain-injured "amnesic" subject is handicapped when required to check or inhibit some dominant or alternative response tendency. Thus, the rather vague "memory deficit" construct becomes much more concrete when the deficit is viewed as being attributable to interfering response tendencies. This simplifying conceptualization has

[1] This research was supported by Canadian Research Council Grant AP-65 and by National Institute of Mental Health Grant MH 08016.

[2] See Chapter 2, this volume.

already contributed to an integration of human and animal research. A different type of methodology which will undoubtedly lead to additional clarifying experiments has been demonstrated.

A similar "response interference" conclusion, initially unanticipated, was arrived at several years ago during the Warren and Akert (1964) symposium. It was by then well known that animals with a frontal lesion had difficulty in locating a reward which just previously and in their full view had been placed under one of two identical test objects. There were also some inconsistencies, not unlike those previously described with respect to hippocampal-temporal lesion cases. The human frontal handicaps were rather subtle and appeared to differ from the deficits demonstrated by animal subjects. The conventional description, "a deficit in immediate memory," turned out to be inadequate when tested with more specific validating procedures. During the symposium, a variety of experiments using a number of traditional and innovative procedures with rat, cat, squirrel monkey, rhesus monkey, and human subjects were carefully reviewed. A common feature seemed to underlie most of the data. The frontal animals had had unusual difficulty in inhibiting a dominant response repertoire. Much of the frontal data and the hippocampal-inferior temporal data reviewed by Professor Weiskrantz suggest that diminished "behavioral flexibility" may be attributable to a variety of lesions. This does not imply functional equivalence of different cortical areas. There are considerable data showing that different lesion locations are correlated with differential "flexibility" losses (Grossman, 1967). Further lesion research will undoubtedly reveal additional general and specific "flexibility" deficits.

The memory experiment by Dr. Jarrard[3] demonstrates that verbal memory procedures developed for use with human subjects can be simulated so as to be applicable to primates and, very likely, to other animals. His successful development of automated objective procedures for investigating STM in stumptail macaque subjects holds considerable promise. The general procedure is versatile and can be adapted for the investigation of many of the parameters that have yielded systematic information in the research area of human STM. The described program has the potential of obtaining significant data in the relatively neglected research area of animal memory. In addition to the possibility of delineating differences and similarities between human and animal memory functions, the more precise control which can be achieved with animal subjects could provide the basis for the construction of more precise LTM and STM models. Dr. Jarrard's program provides a good example of his described process theory approach. A series of interrelated experiments has been conducted which could lead to a descriptive model or formulation representative of STM in primate subjects.

[3] See Chapter 1, this volume.

The following memory research program illustrates the alternative strategy which Dr. Jarrard described—a comparative approach in which an attempt was made to obtain reliable differences among various species on similar behavioral tasks. The basic problem consisted of a modification of the classical delayed-response procedure. Initially, the delayed-response problem was designed to determine if animals could react on the basis of "mental imagery." The typical procedure involved confronting the subject with a number (usually two) of identical test objects. A momentary spatial cue signifying the locus of reinforcement was presented. After a brief lapse of time, the subject was permitted to select one of the identical test objects. Because the cue signifying reward was absent at the time of selection, "correct" (i.e., reinforced) responses implied the existence of a memory trace, or "image." Although the issue of mental imagery did not outlive the trend toward objectivity, the delayed-response procedure has survived because it produced behavioral results of general interest. When human and animal performance levels are compared, it becomes obvious that this apparently simple task is an extremely difficult one for animal subjects, even when they are very well trained. As was mentioned previously, it is now well established that frontal subjects from a variety of species exhibit a severe delayed-response performance deficit (Warren & Akert, 1964).

The delayed-response procedure has been used with relative success as a means of estimating the comparative learning ability of different species (Meyers, McQuiston, & Miles, 1962). It is believed that a direct comparative psychology of learning provides supplementary information which contributes to a comprehensive understanding of "intellectual" functions. It should be mentioned that the establishment of meaningful comparative learning ability trends among different animal species has proven to be rather difficult. Except for the delayed-response and learning-set problems, conventional learning procedures have not demonstrated systematic phylogenetic ordering of species (Warren, 1965).

The comparative research program to be described in the following paragraphs was initiated with the standardized delayed-response procedure using the Wisconsin General Test Apparatus (WGTA). This commonly used apparatus consists of two compartments, one to house the animal during testing and the other equipped with a movable tray having two foodwells spaced about 12 in. apart near the front edge. A row of vertical bars and a movable opaque screen separate the two compartments. Each delayed-response trial consisted of the following sequence of events: The experimenter raised the opaque screen, extended a morsel of food over the tray, and conspicuously placed the food into one of the two foodwells. He then simultaneously covered both foodwells with identical test objects and started a timer with a foot switch. Upon completion of the designated delay period, the tray was pushed forward and the subject was permitted to displace one of the objects. If the selection was correct, the subject obtained food reinforcement; incorrect selections were followed by immediate withdrawal of the test tray.

This simple delayed-response procedure was used with a number of different species as a comparative learning test. Every effort was made to circumvent the usual methodological shortcomings of comparative research by maintaining an identical test procedure with all species, using preferred rewards, trying to assure that no subject group had any obvious advantage due to a particular type of sensory-motor development, and obtaining fairly extensive behavioral samples after thorough adaptation and pretraining. These precautions should have minimized the possible differential emotional interfering reactions that were described by Professor Harlow.[4]

The first experiment in the series compared delayed-response performances of the rhesus monkey and the marmoset (Miles, 1957). The marmoset (of small size, having clawlike nails and a relatively smooth cortex) was considered to be representative of the lower phylogenetic ranks within the primate order. The rhesus macaque, which has long been a popular laboratory subject, was viewed as having fairly advanced phylogenetic rank among the primates. To assure that training experiences were the same over an equivalent number of trials, sequences were planned so that each subject experienced an equal number of trials under all delay conditions within each daily test session. Another experiment (House & Zeaman, 1961) investigated the delayed-response learning of 33 mentally defective children who ranged in mental age from 2 to 5 years. A group of normal 5-year-old children ($N = 5$) was tested by Miles. Delayed-response learning by domestic cats was studied by Meyers et al. (1962). Although the actual number of training trials differed somewhat among groups, all groups except the group of normal human children (who experienced 250 trials) received sufficient training to approach asymptotic performance levels.

Figure 1 illustrates performances of the subjects in the four experiments. Comparisons among the five species show that performances differed significantly and were consistent with estimations of phylogenetic rank. The performance of young normal human children, which was highly superior, was followed by that of human mental retardates. This group outperformed rhesus monkeys, who were clearly superior to the marmosets, and the marmosets in turn were better than domestic cats.

The delayed-response procedure has also been used to compare learning ability at different age levels within a species. Rhesus monkeys at 60, 90, 120, and 150 days of age and as adults were tested by Harlow, Harlow, Rueping, and Mason, (1960). The results of this investigation indicated a general and reliable improvement in performance as a function of age. These comparative and developmental performance differences, and the marked performance deficit re-

---

[4] See Chapter 5 and 6, this volume.

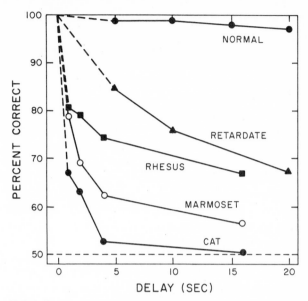

FIG. 1. Percent correct responses as a function of delay interval.

lated to frontal lesions, all suggest that the delayed-response task may represent a distinctive, perhaps a basic learning process.

Although the conventional delayed-response procedure has served as a useful comparative test of learning ability, the upper difficulty range of the problem is so limited that the technique cannot be used to investigate the relative learning ability of normal human children and adults. Even the 5-year-old human subjects made so few errors that the data showed virtually no within-species behavioral differentiation.

A simple technique planned as a means of extending the difficulty range of the delayed-response problem involved varying both length of delay and number of response alternatives. A row of identical 3/4-in. sq cubes spaced 1/2-in. apart were placed in a line 1 in. from the front of the test tray. Each cube covered a small foodwell. At the beginning of each trial, the cube covering the "correct" locus was placed 1 in. behind the designated foodwell. The experimenter raised the opaque screen, held a morsel of food over the middle of the test tray, paused momentarily, then conspicuously placed the reinforcement in the uncovered foodwell and positioned the test object. At that moment the illumination was turned off and timing of the delay interval began. Intervals were measured by means of a metronome which emitted a low intensity "clack" once every second. At termination of the designated delay, illumination was turned on as the test tray was being pushed forward so that the subject could make an object selection. A response was defined as touching one of the test objects; as soon as an object was touched, the experimenter quickly retracted it by pulling on a small attached chain. A correct selection uncovered the available reinforcement; an incorrect selection was followed by immediate withdrawal of the test tray. During a lengthy series of test trials, each subject experienced progressively longer delay intervals combined with an increasing number of response alternatives. In effect, the subject was repeatedly being presented with a single-row spatial location task comprising a varying number of equally spaced response alternatives.

Extensive training was required to obtain a stable estimate of upper capacity limits. It was necessary to increase the difficulty level very cautiously. Even a short sequence of errors disrupted the performance level, apparently because of partial extinction of behaviors (observing reward placement, waiting calmly during the delay, making an immediate choice when the tray was advanced, etc.) essential for successful performance.

Each subject first experienced an adaptation period of 100 trials under a two-choice, 3-sec delay condition. Then the subject received further testing with delays of 3, 5, 7, or 10 sec and with number of response alternatives gradually increasing. Performance was judged sufficient to warrant changing to the next difficulty level—either longer delay or additional alternatives—if only 2 errors occurred within 20 trials or 6 errors occurred within 30 trials. If correct selections dropped to 14 or less within 20 trials, difficulty level was descreased until the criterion for increasing difficulty was achieved. To assure that the subjects remained well trained, a series of 5 trials with delays of 1 and 3 sec began each daily test session. Location of the "correct" object was determined by a restricted random series so that each position was rewarded equally often and no given position was rewarded more than twice consecutively. This testing technique is akin to the psychophysical method of limits and is even more tedious because the series is explored stepwise only in an upward direction.

This modified delayed-response procedure was first tried out with rhesus monkeys and was then more extensively used with four squirrel monkey subjects. Figure 2 illustrates the test situation, with a subject facing a seven-alternative problem following reward placement.

The effects of length of delay and number of alternatives are shown in Fig. 3. Each point shows percent correct over 100 trials after about 3000 training trials. The independent variables had a very significant effect upon performance. Results of an analysis of variance revealed that effects of both variables were significant beyond the 0.01 level of confidence. When confronted with the conventional two-choice problem, the subjects could "hold" the event of reward placement with virtually no loss throughout a 10-sec period of delay. A noticeable increase in difficulty level was evident as the number of response alternatives increased from 2 to 3 to 5 to 7.

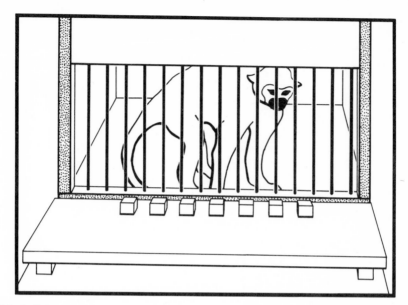

FIG. 2. Delayed-response test arrangement showing a squirrel monkey confronted with a 7-alternative display.

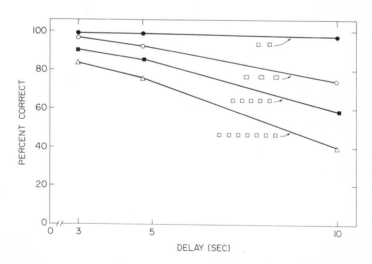

FIG. 3. Percent correct responses as a function of number of alternatives and length of delay.

In subsequent research the same procedure was used to assess the comparative delayed-response performance of squirrel monkeys, cats, marmosets, and rhesus monkeys. Human children and rat subjects also were tested but with a modified procedure. With the exception of the rats, all animal species received 300 two-choice, 3-sec-delay pretraining trials followed by thorough testing with the previously described procedure on problems consisting of 3- and 5-sec delays and five alternatives. All animal subjects experienced from 3500 to 4000 test trials. Because the children (whose average age ranged from 58 to 64 months) quickly caught on to the rules of the game, they received only 25 pretraining and 100 test trials. Following the prescribed test experience, all subjects, including rats and children, experienced an additional 200 trials with delays of 3 and 5 sec on a five-alternative problem.

A heroic attempt was made to train four rat subjects. After 500 trials with very little apparent success, a modified procedure was developed. For the rat subjects the apparatus consisted of a 12-in. wide $\times$ 12-in. high $\times$ 18-in. long enclosure with a glass partition facing a test tray. The test tray was painted flat black with a 1/2-in. sq white patch surrounding each of five foodwells which were spaced 2 in. apart. (If the test objects were closer together, it was not possible to obtain a discrete response from rat subjects.) The 3/4-in. sq objects were also painted white. On each test trial the object was removed from one foodwell and placed behind it, and that location was brightly illuminated by a spot of light 1 in. in diameter. A 97-mg food pellet was shown to the subject and placed in the illuminated foodwell, whereupon the foodwell was covered with the test object. Timing of the delay interval was initiated by the experimenter's tapping the "baited" object twice with his index finger. During the delay interval the spot of light was turned off but dim room illumination (a 25-W bulb) remained on. At the end of the delay interval the glass restraining partition was raised and the subject was permitted to displace one of the test objects. A correct selection was reinforced; the tray was withdrawn immediately after an incorrect choice.

Despite 500 extra trials, the additional cues and more widely separated test objects, the performance of the rats was decidedly inferior to that of the other subjects. Levels of performance of the various subject groups are shown in Fig. 4. These data showed reliable differences among the species that were generally consistent with phylogenetic rank: Proficiency levels increased from rats to cats to marmosets and squirrel monkeys and then to rhesus macaques, with human children clearly outperforming all other groups. There was no significant difference between marmosets and squirrel monkeys, but all other differences between groups under the 5-sec delay condition reached the 0.01 confidence level.

Another effort was made to obtain a relatively standard comparison of the delayed-response capacity of several subject groups. The problem for animals and children consisted of five alternatives combined with various lengths of

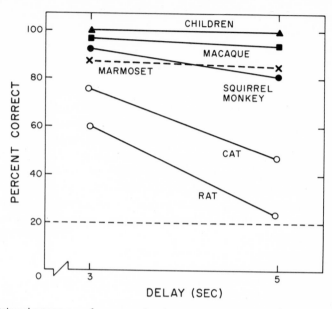

FIG. 4. Delayed-response performance of various species tested on a 5-alternative problem.

delay. Human adults were tested under a 36-choice, 60-sec delay condition. After all subjects had experienced testing as previously described (about 4000 trials for the animal groups, 200 trials for the children, and 100 trials for the adults), each group received additional trials as follows: animal subjects, 200-400 trials; children, 200 trials; adults, 25 trials. In this series of experiments the previously described supplementary cues (spot of light and tapping of the test object) were not used with the rat subjects.

Comparative performances are illustrated in Fig. 5. Among the animal subjects, the definite superiority of the rhesus monkeys and the very low performance level of the rats are quite obvious. Children were decidedly superior to all animal groups. It is clear that all other groups were completely outclassed by the human adults even though they were faced with a 36-alternative problem in combination with a delay of 60 sec.

It is apparent that the proficiency levels of human children and adults were close to the maximum possible (see Figs. 4 and 5). A subsequent series of experiments was planned to obtain a more accurate estimate of the delayed-response capacity of young and adult humans. Both number of alternatives and length of delays had to be considerably increased. Although the test procedure was similar to that used with animal subjects, some modifications were required in order to produce a significant number of errors.

The experimental arrangement for the children consisted of a 6-1/2-ft. test board placed across two chairs or upon a low table. A row of 3/4-in. sq cubes with sides 1/2-in. apart were positioned 1 in. from the edge of the test board facing the subject. The subject and the experimenter sat on opposite sides of the board. Illumination was provided by a dim (7-W) minilamp placed on the floor 1 ft behind the test board and projecting away from the subject and downward toward the floor. (A low illumination intensity was necessary in order to assure that the problem remained one of spatial location; with normal room lighting, the subjects could detect subtle differences among the "identical" test cubes.)

To begin each test trial the experimenter pointed to and touched one of the cubes with a cylindrical pointer (of 1/4-in. diameter and 1-1/2 ft. in length). The pointer was then withdrawn, the lamp turned off, and the designated delay interval timed by counting along with the 1-sec "clacks" of a metronome. At the end of the interval the lamp was turned on, and the subject was asked to indicate (with a 3-ft pointer) the cube that had previously been touched by the experimenter. Instructions to the subject were as follows:

> "We wish to find out how accurate you are in locating a given position following a time delay. Although the task appears to be an easy one, some errors are made by most people. Please try to be as accurate as possible.
>
> "I will point to one of the cubes, then turn out the light. After a time delay, the light will be turned on. Then you are to indicate which block was previously pointed to with the pointer."

Immediately after a selection was made, the subject was informed whether his choice was correct or incorrect; if incorrect, the correct alternative was not indicated.

Each group of children, consisting of six subjects whose most recent birthdays were within six months of the designated age level, received 300 test trials. Depending upon age level and preference, the children were rewarded with either candy, toy charms, or pennies at the end of each daily test session of 60 trials. In addition, an adult group consisting of six college students was tested for 180 trials. All subjects experienced a 64-alternative condition combined with delays of 1, 5, 10, or 30 sec. The adult, 12-year-old, and 8-year-old groups were also tested with longer delays of 15 min and 24 hr. Under these longer delay conditions the subject left the room during the delay period and returned about 2 min before termination of the interval. After the subject was reseated, the lamp was turned off until the end of the designated delay, whereupon it was turned on and the subject was permitted to make an object selection. It should be mentioned that all subjects did not experience equivalent trials under all delay conditions. Since it was obvious that the younger subjects could not cope with the longer delays, testing was concentrated within the range of difficulty where few errors occurred.

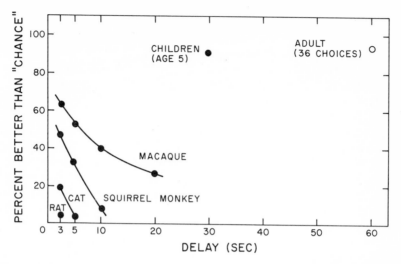

FIG. 5. Delayed-response performance of various species. Adult human subjects experienced a 36-alternative problem; all other species were tested with 5 alternatives.

Results of this series of experiments are shown in Fig. 6. It is apparent that both age level and length of delay had noticeable effects upon performance. All within-subject comparisons between the 1- and 30-sec conditions were significant beyond the 0.01 confidence level. Most of the comparisons between subject groups of different ages also reached the 0.01 level (exceptions are indicated in Fig. 6). This positive relationship between age and delayed-response performance has also been demonstrated in rhesus macaques (Harlow *et al.*, 1960).

A final effort was made to obtain a more accurate approximation of human adult capacity. The procedure was similar to that previously used with children of different ages except that number of alternatives (up to 128) was markedly increased. The test objects were aligned 1/2 in. apart along the front edge of a long (13-ft) test board which was placed 3 ft from a blank wall. The minilamp was placed on the floor 1 ft behind the test board so that light was projected downward toward the floor and wall, providing very dim and indirect illumination of the test area. The subject was seated 5 ft directly back from the middle of the row of test cubes, and the experimenter was seated on the opposite side of the board. At the beginning of each test trial the experimenter briefly indicated one of the objects with a pointer, whereupon illumination was turned off and timing of the delay interval began. At the end of the delay period the lamp was turned on and the subject indicated one of the objects with his pointer. If the delay interval was 15 min or longer, the subject left the room and returned about 2 min before termination of the interval. After the subject was reseated, the trial proceeded as previously described.

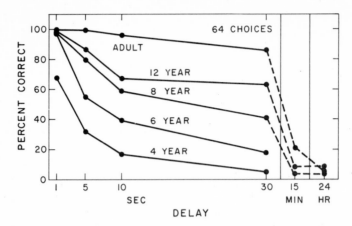

FIG. 6. Percent correct responses as a function of age level and length of delay.

Because a large number of errors occurred under the longer delay conditions, the conventional correct-incorrect dichotomy did not appear to be a suitable performance classification. In contrast to the animal groups, human subjects continued performing and using appropriate test strategy even after making many errors. Almost all human errors were not chance selections but occurred in the vicinity of the designated test object, i.e., were "near misses."

By using information theory it was possible to estimate the informational value of incorrect choices. Information theory was used merely as a descriptive tool without implying any theorectical orientation. A trained experimental subject was viewed as a "transmitter" of spatial location information. Both the delayed-response trial setting arrangement (number of alternatives) and the subsequent response distribution of the subject were transformed into the conventional informational unit, the binary digit or "bit": log to the base 2 of the number of alternatives — H (information in bits) = log M (number of equiprobable alternatives). By using a multivariate technique developed by Garner and Hake (1951) and by McGill (1954), it was possible to delineate the following informational components: $H(x)$, the estimated stimulus "input information" (number of spatial location alternatives or test objects); $H(y)$, the estimated response information (from a tabulation of object positions selected by the subject); and $H(xy)$, consisting of the joint occurrence of the designated alternative and the subject's object selection (correct responses). It was also possible to estimate the amount of shared or transmitted information, T. $T(xy)$ = $H(x) + H(y) - H(xy)$. Since all subjects were familiar with the procedure and an object selection was almost always related to the one designated by the experimenter, $H(x) + H(y)$ was consistently greater than $H(xy)$ (Attneave, 1959).

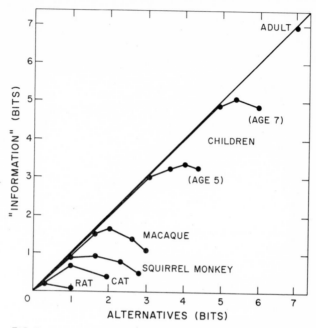

FIG. 7. Amount of "transmitted" spatial location information.

Figure 7 illustrates the results of an exploratory experiment designed to compare the delayed-response "information transmitting" capacities of various groups of well-trained subjects. Each data point for the animals and for the 5-year-old humans is based on an additional 50 test trials with the subjects that had experienced the testing series referred to in Figs. 4 and 5. The 7-year-old children and the adults received 200 and 100 training trials, respectively, followed by the 50 test trials summarized in Fig. 7. During this 50-trial series the delay interval was constant (10 sec) and the number of alternatives progressively increased. Amount of transmitted information increased as a function of either phylogenetic rank or age. The rat subjects performed at about chance level, whereas the human adults transmitted seven bits (128 alternatives) of spatial location information almost perfectly. It should be noted that an increase in the number of alternatives typically resulted in a proportionate increase in errors, so that the amount of information transmitted remained about the same. This result is consistent with the idea of a limited discriminative "channel capacity" as described by Miller (1953).

Although there is no doubt that the performance of adult humans outclassed that of all other groups, the level of difficulty was still too low for making even a crude estimation of their upper capacity level. Another experiment therefore

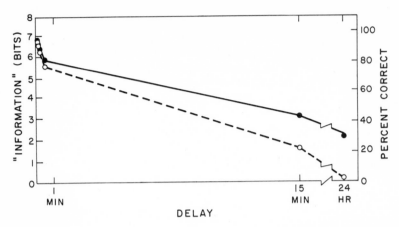

FIG. 8. Delayed-response performance levels presented as percent correct responses and as amount of "transmitted" information.

was designed to test five adults (advanced undergraduate and graduate students) with a procedure which combined 128 alternatives with an extensive range of delays (1, 5, 10, 30, and 60 sec; 15 min; and 24 hr). Figure 8 summarizes the performances of the subjects on 20 to 50 trials under each delay condition. Although the percentage of correct object selections became very low as length of delay increased, the subjects still were transmitting a significant amount of spatial location information. Even after the 24-hr delay, subjects were able to recall the approximate location of the designated stimulus object.

These data and the previously described human delayed-response performances indicate that the comparative status of *Homo sapiens* is indeed outstanding. As illustrated previously in Fig. 5, the performance of the best animal group (rhesus monkeys) was only about 30% better than chance on a five-alternative, 20-sec delay problem, whereas adult human subjects performed almost perfectly when a delay of 60 sec was combined with 36 alternatives. There obviously is a very great difference in capacity between monkey and man. Although the well-trained animal subjects experienced an easier version of the delayed-response problem (the subject was closer to the test tray and viewed larger objects spaced farther apart), it appears that even an infinite extension of training would not have produced equivalent performance. Animal subjects clearly could not match human performance on the 128-alternative, 86,400-sec delay problem.

It should be pointed out that the delayed-response problem does not provide any apparent special advantage for the human subject. The test procedure is representative of the S-R "noncognitive" type of learning described by Meyer.[5]

[5] See Chapter 4, this volume.

The elaborate linguistic ability of man would not appear to be an obvious asset in the soulution of the simple delayed-response problem. Because selections of stimulus objects near either end of the test board were excluded, numerical habits or coding were probably used minimally, if at all. Thus, human as well as animal subjects were required to make a simple indicative response following presentation of a spatial location cue. Despite being stripped of his traditional problem-solving advantages and being presented with a nonchallenging, almost degrading, task, the human subject completely outclassed the best animal performer. It is suggested that this delayed-response performance difference between humans and animals is of sufficient magnitude to be described as qualitative rather than quantitative. Perhaps some of us will be reassured by this finding!

REFERENCES

Attneave, F. *Applications of information theory to psychology: A summary of basic concepts, methods, and results.* New York: Holt, Rinehart & Winsten, 1959.

Garner, W. R., & Hake, H. W. The amount of information in absolute judgments. *Psychological Review*, 1951, **58**, 446-459.

Grossman, S. P. *A textbook of physiological psychology.* New York: Wiley, 1967.

Harlow, H. F., Harlow, M. K., Rueping, R. R., & Mason, W. A. Performance of infant rhesus monkeys on discrimination learning, delayed response, and discrimination learning set. *Journal of Comparative and Physiological Psychology*, 1960, **53**, 113-121.

House, B. J., & Zeaman, D. Effects of practice on the delayed response of retardates. *Journal of Comparative and Physiological Psychology*, 1961, **54**, 255-260.

McGill, W. J. Multivariate information transmission. *Psychometrika*, 1954, **19**, 97-116.

Meyers, W. J., McQuiston, M. D., & Miles, R. C. Delayed-response and learning-set performance of cats. *Journal of Comparative and Physiological Psychology*, 1962, **55**, 515-517.

Miles, R. C. Delayed-response learning in the marmoset and the macaque. *Journal of Comparative and Physiological Psychology*, 1957, **50**, 352-355.

Miller, G. A. What is information measurement? *American Psychologist*, 1953, **8**, 3-11.

Warren, J. M. Primate learning in comparative perspective. In A. M. Schrier, H. F. Harlow, & F. Stollnitz (Eds.), *Behavior of nonhuman primates.* Vol. 1. New York: Academic Press, 1965.

Warren, J. M., & Akert, K. (Eds.) *The frontal granular cortex and behavior.* New York: McGraw-Hill, 1964.

Author Index

Numbers in italics refer to the pages on which the complete references are listed.

Subject Index